AF443090

Physics of Solids, Nuclei and Particles

Physics of Solids, Nuclei and Particles

Editor

R. Sahu

Narosa Publishing House

New Delhi   Chennai   Mumbai   Kolkata

Editor
R. Sahu
Department of Physics
Berhampur University
Bhanja Bihar, Berhampur
Orissa, India

NAROSA PUBLISHING HOUSE PVT. LTD.

22 Daryaganj, Delhi Medical Association Road, New Delhi 110 002
35-36 Greams Road, Thousand Lights, Chennai 600 006
306 Shiv Centre, D.B.C. Sector 17, K.U. Bazar P.O., Navi Mumbai 400 703
2F-2G Shivam Chambers, 53 Syed Amir Ali Avenue, Kolkata 700 019

www.narosa.com

Printed from the camera-ready copy provided by the Editor

ISBN 81-7319-641-9

Published by N.K. Mehra for Narosa Publishing House Pvt. Ltd., 22 Daryaganj, Delhi Medical Association Road, New Delhi 110 002 and printed at Rajkamal Electric Press, New Delhi 110 033, India

Preface

Over the years the developments in different fields of Physics have been so rapid and so diversified that many times physicists working in a particular branch are not aware of the developments in other fields of research. However, in most cases, the underlying physics behind the different fields are essentially same. For example, pairing between identical particles was first introduced in nuclear physics by Mayer and Jensen to provide a description of the magic nuclei. This idea was then used by Bardeen, Cooper and Schiffer to describe the superconductivity phenomena in condensed matter physics. This theory, popularly called BCS theory, goes far beyond the original idea and finds extensive use in structure studies in finite nuclei in addition to the condensed matter physics. Again the theory of super symmetry is a highly developed subject in high-energy physics. In a recent experiment by Metz et al. (Phys. Rev. Lett. 83,1542(1999)) strong evidence for the existence of super symmetry in atomic nucleus have been found. Hence it was felt that experts in different branches of physics should interact in a seminar. This was the motivation for starting the National seminars on "Recent Advances in Physics" by Berhampur University every two years. The main encouragement for holding the National Seminar in September 2003 was given by Prof. S.N. Behera who was then the Vice Chancellor of our university and Prof. R.K. Choudhury, Director, Institute of Physics, Bhubaneswar. The present book is the outcome of the lectures given by eminent physicists in the fields of Condensed Matter Physics, Nuclear and Particle Physics.

The book contains six articles in condensed matter physics (including two in electronic science), one in quantum mechanics and three in nuclear and particle physics. The topics can be easily understood by the graduate students in any of the above fields. The articles also contain the original research contributions of the authors. I hope, the book will be useful to the readers.

I am thankful to the members of the organizing committee who have spent their valuable time in making the seminar a success. I am indebted to UGC and the Director, Institute of Physics for liberal financial support. I am thankful to Prof. S.N. Behera and Prof. R.K. Choudhury for their guidance and continuous support in organizing the seminar. I am also thankful to the faculty and the students of Physics Department in making the seminar a success.

R. Sahu

Contents

Physics of Solids, Nuclei and Particles
Editor: R. Sahu

Some Recent Developments in Magnetism

S.N. Behera

Institute of Physics, Sachivalaya Marg, Bhubaneswar 751 005, India

1   Introduction

The subject of magnetism is a very old one and it has been studied for almost 400 years. Even though the first recorded studies
of the properties of magnets were by Gilbert in the 17th Century;
magnets and their properties seems to have been known to the Chi-
nese and Indians for almost 1000 years [1]. During the early days
of the development of physics, the ease with which magnets and
magnetic fields were available in the laboratories, made it a popu-
lar probe for studying its effect on other physical phenomena and
material properties. Among many properties of magnets and the
effects of magnetic fields that we have become familiar with, the
galvanomagnetic effects were one of the first to be studied and are

the most interesting ones. For example, it was realized that a current carrying loop produces a magnetic field. Alternately, a time varying magnetic flux through a circular wire loop can produce an electro-motive force (emf.); an effect which came to be known as the Faraday's laws of induction. These two effects clearly indicates that electricity and magnetism are phenomena related to each other. Later this led to the unification of the two phenomena into electro-magnetism by James Clark Maxwell.

An important galvanomagnetic effect which has been studied in some detail is the change of electrical resistivity of a conductor in a magnetic field, which is called magneto-resistance. In 1856, Lord Kelvin discovered that the resistance of a metal increases with increasing magnetic field, which is called positive magneto-resistance. However the observed magnetoresistance (MR) in a metal is usually very small, being less than a small fraction of 1%. What is the origin of MR and why is it so small? The effect arises because a charged (q) particle; moving with a velocity $\vec{V}$ in a particular direction, experiences the Lorentz force in a magnetic field $(\vec{B})$; which is given by $\bar{F} = (q/c)\vec{V} \times \bar{B}$, c being the velocity of light. The direction of this force being parpendicular to both $\vec{V}$ and $\bar{B}$ makes the charged particle (e.g., electrons) move in a helical path. This deviation of the particle from its original path, tends to increase the electrical resistance. It is also well known that because of this deviation in the path of the charged particle, a voltage developes in the direction of the force known as the Hall voltage. Because of this voltage, in the steady state a current is set up which tends to cancel the applied magnetic field. As a consequence once the stady state is reached the electrons in the metal do not experience the Lorentz force any more, there by resulting in a small value of the MR. This being purely a geometrical effect, by choosing appropriate geometries, it is possible to enhance the positive MR, and this enhanced MR is called the Extraordinary Magnetoresistance (EMR). Yet another familiar effect associated with the geometric changes in the volume of some magnetic material during magnetization is called magnetostriction, which is a mechanical effect. For example, when a single crystal of iron is magnetized, its length in the direction of magnetization increases. This results in the anisotropy of the magnetized material which in turn can affect its optical properties. In contrast some of the optical properties also get affected in the presence of magnetization, even in the absence of magnetostriction.

The most well known magneto-optic effects are the Faraday and the Kerr effects. When a plane polarized light is passed through an isotropic material of high refractive index, the direction of the electrical field vector of the incident light is found to rotate if the substance is placed in a strong magnetic field. This is known as Faraday effect. It was also found that when a plane polarized light was reflected normally from a polished pole piece of an electromagnet, the reflected light becomes elliptically polarized. The later is known as the Kerr magneto-optic effect. Kerr showed that the magnetic field gives rise to a vibration of the electric field vector, which is known as Kerr component, perpendicular to the direction of vibration in the incident light. The Kerr reflection in turn can be used to measure the magnetization of a material.

2   Magnetic thin films

So far what is discussed concerned with how some physical phenomena get affected when the material is either subjected to a magnetic field or magnetized. However, in no way this throws any light on the origin of magnetism. A proper understanding of magnetism had to wait till after the development of the quantum theory. Therefore even though magnetism is an old subject its understanding is relatively recent. Magnetism owes its existence entirely to quantum mechanics. According to our present knowledge, one requires quantum mechanics to understand the origin of magnetism at the level of an atomic scale. The magnetic moment of a free atom has three principal sources : (1) the spin with which electrons are endowed (2) their orbital angular momentum about the nucleus and (3) the change in the orbital moment induced by an applied magnetic field. The spin of the electron is a natural consequence of the relativistic theory or the Dirac equation. Considering the fact that a proper understanding of the origin of magnetism was achieved only in the middle of the last century, it is no wonder that new effects are still emerging in the field of magnetism.

In recent years the study of the magnetic effects in low dimensional systems in general and layered magnetic structures in particular have made rapid progress due to their large scale applications in technologies related to magnetic sensors and magnetic information storage, with emerging requirements for non-volatile computer memory, integrated microwave circuits and optical communications etc. Thin films of magnetic materials have been of particular interest in this regard. In 1884, Kundt first studied the properties of thin magnetic films. In relatively more recent past, scientists tried to search for soft magnetic materials, those which can be easily magnetized and in which the direction of magnetization is easily reversible, for storage of information in computers. The commonly used material in technology for making magnetic thin films is the compound permalloy which is an alloy of nickel and iron, whose chemical composition is $Ni_{0.8}Fe_{0.2}$. Permalloy possesses the desired properties namely an easy axis of magnetization, i.e., an axis along which the material is most easily magnetized.

It is because of the demands of technology that much progress is achieved in the study of the properties of magnetic thin films and multilayers. In turn this resulted in the discovery of many new phenomena like dimensional cross over, inter layer exchange coupling (IEC) exchange bias (EB) and giant magnetoresistance (GMR). The physics behind some of these phenomena will be discussed in what follow.

A magnetic thin film may be deposited on a suitable substrate by the state of the art methods like Molecular Beam Epitaxy, (MB),

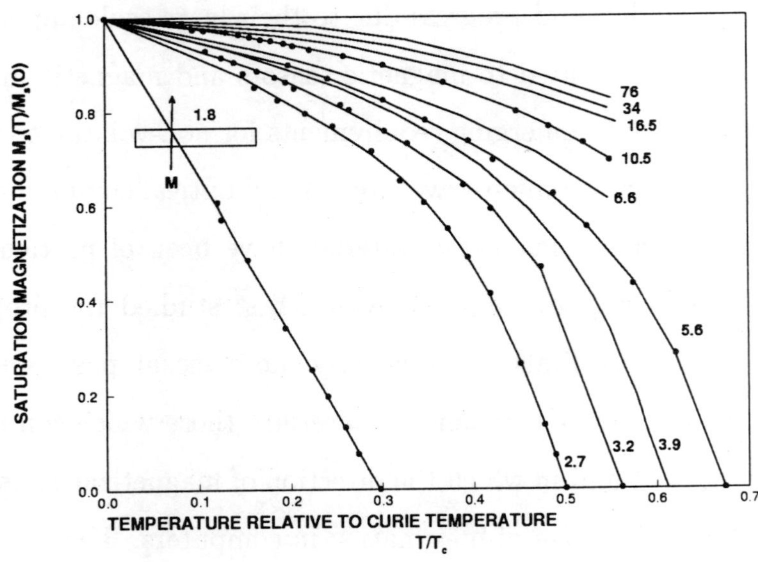

Figure 1: Temperature dependence of saturation magnetization in thin films of $Ni_{48}Fe_{52}$ alloy. (taken from P. GrüBerg [2].)

as well as Metal Organic Chemical Vapor Deposition (MOVED) and by several other methods. These methods envisage at producing the thin film atomic layer by atomic layer. As a result if films of two different materials are grown on top of one another, the interfaces tend to be perfect. It is interesting to ask how the magnetic properties of the material change with decreasing film thickness. One such property which can be looked at is the change in the critical temperature T_c above which a thin film loses magnetic ordering as a function of the film thickness. As the film is gradually made thinner, it is found that the value of T_c decreases with the thickness, i.e., with approaching two-dimensionality. The variation of the saturation magnetization $M_s(T)/M_s(0)$ with T/T_c, T_c being the Curie temperature for the material in bulk, is found to exhibit the behaviour shown in Fig. 1, and becomes more and more linear as the thickness of the film is reduced. The numbers in the figure indicate the thickness of the film in monolayers. The direction of magnetization remains in the plane of the film for decreasing thickness till one reaches very small thickness typically of magnitude, say, 1.8 monolayers (one monolayer being about $3\mathring{A}$ thick) for a film of $Ni_{48}Fe_{52}$. At this stage the magnetization becomes perpendicular to the plane of the film, which is a surprising result indeed.

3   Exchange Bias (EB)

When a magnetic thin film deposited on a nonmagnetic substrate is cooled, it gets magnetized when the temperature becomes lower than the Curie temperature of the material. In principle when magnetized the direction of magnetization can be any direction in the plane of the thin film, for sufficiently thick films. The fact that there is no preferred direction of magnetization is a consequence of the so called spontaneous symmetry breaking. This is also the reason why magnetic domains get formed. On the other hand when the magnetic thin film is deposited on an antiferromagnetic substrate, and cooled below the Curie temperature it always gets magnetized in a preferred direction. It appears as if there is no spontaneous symmetry braking in this case; which is a surprising result. The system, i.e., the substrate and the magnetic thin film behave as if the substrate provides a biased direction for the magnetization, which is more likely to happen in case of a ferromagnetic substrate. In the later case the presence of a magnetization in a particular direction in the substrate determines the direction of magnetization of the thin film. But in the case of an antiferromagnetic substrate the net magnetization of the substrate is zero, hence it is not understandable, how the substrate provides a bias to the thin film and makes it choose a particular direction for its magnetization. This phenomenon is called "Exchange Bias" (EB), whose understanding from an atomic point of view is not clear. Just as the Interlayer Exchange Coupling (IEC), the phenomenon of Exchange Bias (EB) also remains a puzzle. However the phenomenon finds application in designing devices such as field sensors to control rotating objects [3], simply because EB makes it possible to get the entire thin film uniformly magnetized without producing magnetic domains. As a consequence if one chooses a natural antiferromagnet as the substrate and deposits a magnetic thin layer on top of it with a nonmagnetic metallic layer of appropriate thickness as the sandwiched layer one can produce a synthetic antiferromagnet in which the direction of magnetization in the magnetic layers will be fixed and free of domain structures. Such a synthetic antiferromagnet will provide a much better base for the subsequent deposition of a nonmagnetic and a magnetic layer, the unit as a whole then being used to make the device.

4   Magnetic Multilayers

Magnetic multilayers or metallic magnetic super-lattices are produced by orderly deposition of alternating thin films of two or more metals in which at least one is magnetic. These multilayers exhibit a variety of new and interesting physical effects, which appear in different areas of magnetism like magneto-optics, magneto-statics, magneto-resistance, microwave properties etc. These multilayers and superlattices are also prepared using the same techniques mentioned earlier, for example, evaporation, molecular-beam epitaxy and chemical vapor deposition etc. Several interesting properties of magneto-transport and magnetization have been observed in these multi-layers where the magnetic layers are separated by non-magnetic metallic layers of variable thickness. The magnetization in successive magnetic layers is found to arrange with its direction either parallel or anti-parallel with each other depending on the thickness of the sandwiched non-magnetic metallic layers. For these two arrangements interesting deviations are also observed in their of electrical resistance.

An example of magnetization is schematically presented in Fig. 2 for magnetic metallic multilayers. As can be seen in Fig. 2(a) successive magnetic layers are arranged with their magnetizations anti-parallel while, in Fig. 2(b) the magnetic layers are arranged with

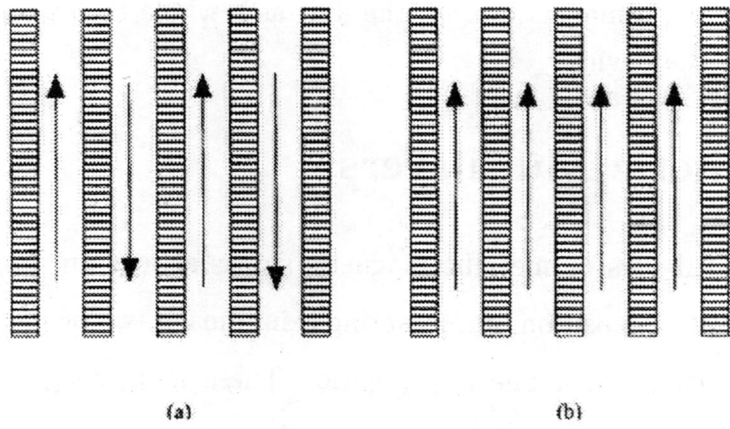

Figure 2: Schematic representations of magnetic multilayers with their direction of magnetizations.

their magnetizations parallel to each other. This parallel magnetization configuration can also result from the application of an external magnetic field to the anti-parallel arrangement in Fig. 2(a). The resistance in the arrangement in which successive layers are antiferromagnetically coupled is found to be large compared to the resistance of the ferromagnetically coupled layers. This behaviour of the resistance in the two magnetic multilayers depicted in For. 2 is quite analogous to the propagation of light in a polarization experiment in which the aligned polarizers allow light to pass through while, the crossed polarizers do not. In case of ferromagnetically coupled layers electrons with their spin orientation parallel to the direction of magnetization can easily pass through and thus the electrical resistance is low but in case of antiferromagnetic alignment the electrons can't get through the structure easily and thus the electrical resistance is high. The fact that by applying an external magnetic field the antiferromagnetic multilayers can be turned into a ferromagnetic one, implies that the electrical resistance of such a multilayer can be lowered by applying a magnetic field. This is negative magnetoresistance which can be large, the relative change in resistance can be upto several per cent of that in the absence of magnetic field. Hence magnetic multilayers are termed as giant magnetoresistance (GMR) systems. Again the decrease in resistance will depend on the thickness of the sandwiched non-magnetic metallic layer as will be discussed below.

5   Interlayer Exchange Coupling (IEC)

Several advanced techniques are used to measure the magnetization of a magnetic multilayer. The most useful methods considering their simplicity and high resolution which are commonly used for this purpose are the scanning electron microscopy with polarization analysis (SEMPA) and the surface magneto-optic Kerr effect (SMOKE).

In these experiments the magnetization of the multilayers is measured as a function of the applied magnetic field which results in the well known hysteresis loop. The characteristics of the measured hysteresis loop provides the information regarding the alignment of the magnetization in the successive magnetic layers of the multilayer. For example if the alignment is ferromagnetic one gets a

                    Physics of solids, nuclei and particles

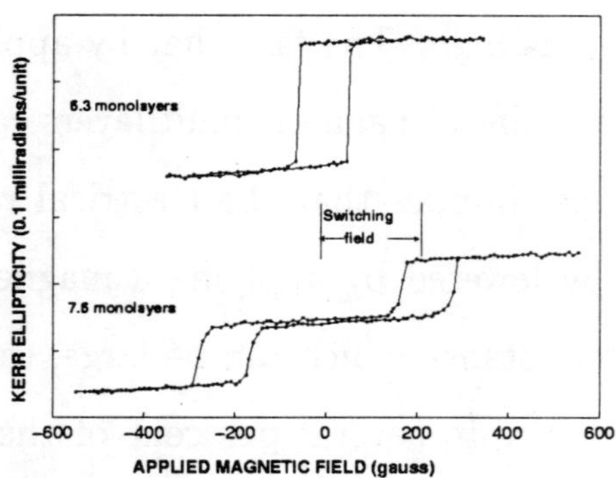

Figure 3: The hysteresis loops of ferromagnetic and anti-ferromagnetic trilayer of $Fe(14)/Mo(\eta)/Fe(14)$. The data were collected by using SMOKE method (taken from L.M. Falicov [3]).

hysteresis loop centred around zero applied magnetic field. In contrast for the antiferromagnetic alignment of the magnetization in consecutive magnetic layers the hysteresis loop will be shifted to a finite value of the applied magnetic field, and there will be a second hysteresis loops corresponding to the reversal of the direction of the magnetic field. The threshold field around which the hysteresis loops appear are called the "Switching fields". An example of this for a magnetic trilayer is shown in Fig. 3. The figure shows the hysteresis loops for a magnetic trilayer of Fe (14 monolayers)/Mo (η monolayers)/Fe (14 monolayers). The thickness of the nonmagnetic Mo layer is η, which varies from sample to sample. Depending upon the average value of η this trilayer shows either ferromagnetic or anti-ferromagnetic alignment of magnetization in the Fe layers as shown in Fig. 3. In this case the average value of η is 6.3 monolayers for the ferromagnetic alignment and 7.6 monolayers for the anti-ferromagnetic alignment of the magnetization in the Fe layers. In the hysteresis loop for the ferromagnetic case, the magnetization starts increasing on increasing the value of the applied field from zero and attains saturation, showing the expected behaviour, because with increasing magnetic field the magnetization in the different domains get aligned in the direction of the applied field. In contrast to this in the antiferromagnetic case first one has to attain a thresh hold value of the applied magnetic field to switch the alignment of the magnetization in all the magnetic layers in the direction of the field, before the field starts aligning the domains; in order to produce the hysteresis loop. Therefore the hysteresis loop will be centred around the switching field, and it will be symmetrically situated on reversing the direction of the field. Thus for the antiferromagnetic case one sees two hysteresis loops.

It is interesting to note that, the alignment of the direction of magnetization in the magnetic layers of the multilayer depends crucially on the thickness of the nonmagnetic layer; as exemplified in Fig. 3. This behaviour is not specific only to the trilayers shown in Fig. 3 but is representative of a general behaviour, where on varying the thickness of the nonmagnetic layer in different samples alternately ferrromagnetic and the antiferromagnetic alignment of the magnetization appears as exemplified in Fig. 4. for various thicknesses of the nonmagnetic Mo layer; as indicated in the figure. While there is ferromagnetic alignment of the magnetization in Fe

layers for the thicknesses of 3.8, 6.3, 9.6, 12.5 and 15.5 monolayers of the sandwiched *Mo* layer, the alignment is antiferromagnetic for the *Mo* layer thicknesses of 5.2, 7.6, 11.2, 13.8 and 17.0 monolayers.

While these measurements were carried out on trilayer samples prepared with different spacer layer thicknesses, a clever way of doing the same experiment at one shot is to prepare a single sample with an wedge shaped non-magnetic layer sandwiched between two magnetic layers. This way the spacer layer thickness varies continuously.

It can also be noticed from fig. 4. that, while there is not much change in the hysteresis loops of the ferromagnetic sandwiches, in the case of the antiferromagnetic trilayers with increasing thickness of the *Mo* layer, the switching field decreases rapidly resulting in the merger of the two hysteresis loops. Therefore beyond a certain critical thickness of the spacer layer, the alignment of the magnetization will remain ferromagnetic. This oscillation in magnetization as a function of the thickness of the spacer layer is clearly demonstrated in the dependence of the switching field as a function of the spacer layer thickness, which is shown in Fig. 5.

This is a totally new phenomenon observed in magnetic multilayers and is called "Interlayer Exchange Coupling" (IEC). This is analogous to the Heisenberg Exchange interaction in magnetic solids. The magnetism in bulk solids is understood by assuming an interaction between neighbouring localized atomic spins(S), which

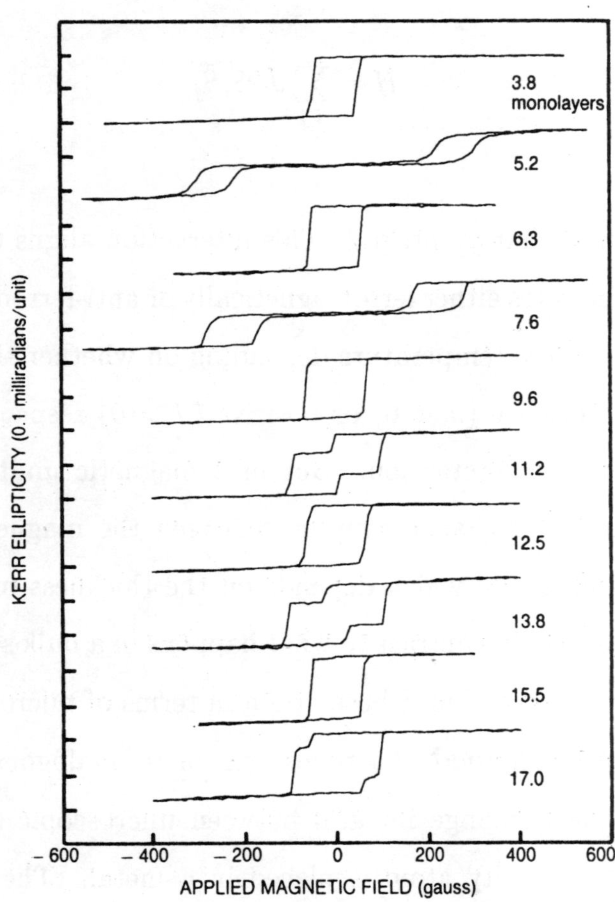

Figure 4: The ferromagnetic and antiferromagnetic alignment of magnetization in the *Fe* layers for various thicknesses of the non-magnetic *Mo* layers. (taken from L.M. Falicov [3]).

is expressed as a Hamiltonian given by

$$H = \sum_{i,j} J_{ij} \vec{S}_i \vec{S}_j$$

J being the exchange integral. This interaction aligns the spins on neighbouring sites either ferromagnetically or anti-ferromagnetically below the critical temperature depending on whether the exchange integral is negative ($J < 0$) or positive ($J > 0$) respectively. This is a microscopic interaction. But in a magnetic multilayer there exists a similar exchange coupling between the magnetizations of neighbouring layers, which depends on the thickness of the spacer layer, i.e., $J(R)$. In contrast to what happens in a bulk solid this is a macroscopic interaction, whose origin in terms of microscopic spins is yet to be understood. There also exists an analogous oscillatory nature of the exchange integral between microscopic spins, when a magnetic impurity atom is placed in a metal. The interaction between the localized spin of the impurity atom and the spins of the itinerant conduction electrons of the metal show an oscillatory behaviour as a function of the distance from the impurity spin. This is the so called Rudderman, Kittel, Kasuya, Yoshida or RKKY interaction. Can one understand the oscillatory nature of the IEC as a function of the thickness of non-magnetic spacer layer from a microscopic point of view? These are questions waiting for answers.

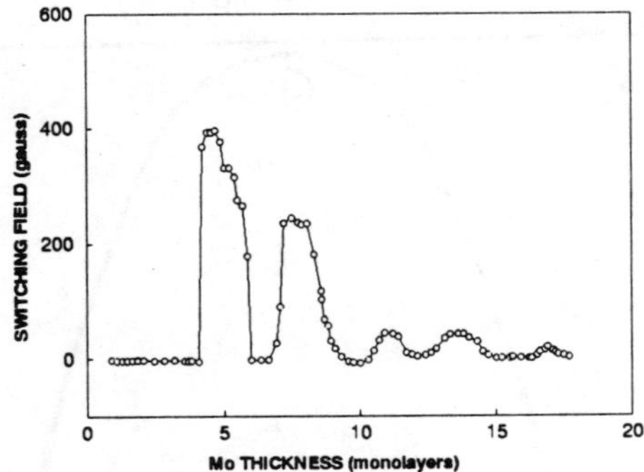

Figure 5: The oscillations observed in the switching field as a function of the thickness of the sandwiched nonmagnetic layer, (taken from L.M. Falicov[3]).

6   Giant magnetoresistance (GMR)

The study of electrical properties e.g. conductivity of the materials in presence of an applied magnetic field provides information about the electronic structure of solids and also the nature of the Fermi surface of metals. The positive magneto-resistance i.e., the increase of resistance in metals upon application of magnetic field gives the convoluted character of the Fermi surface. In contrast the negative magneto-resistance i.e., the decrease of electrical resistance on applying a magnetic field implies increase of orderliness and a reduction of electron scattering. This phenomenon can be explained considering the domain structure inherent in a magnetic material. An applied magnetic field changes the domain structure and reduces the whole material to a single domain. The Bloch walls which separated the magnetic domains and are one of the sources of electron

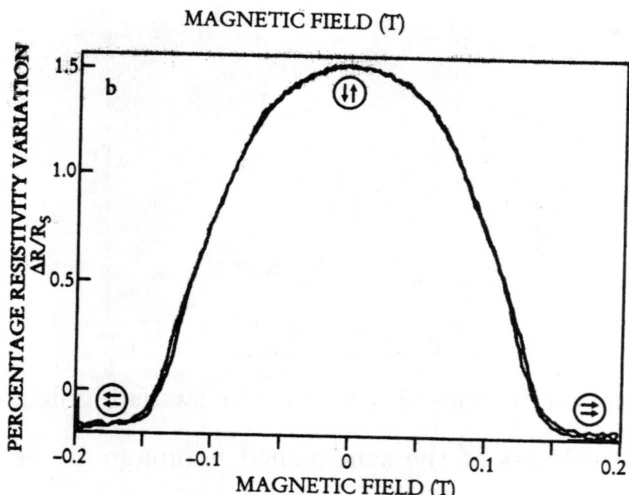

Figure 6: Percentage change of resistivity as a function of magnetic field for the $Fe/Cr/Fe$ trilayer, (taken from P. Grünberg[2]).

scattering are thus eliminated which allows the electrons to transmit easily through the sample giving rise to higher conductivity and a negative magneto-resistance.

The phenomenon of giant magneto-resistance (GMR) is an effect which results in a dramatic variation of the electrical resistivity with the applied magnetic field. It was first observed in magnetic multilayers an example of which is illustrated in Fig. 6. The figure shows the variation of the percentage change in resistivity of an antiferromagnetic $Fe/Cr/Fe$ trilayer from the value R_s which is the resistance of the system at switching magnetic field along the easy axis, plotted as a function of the magnetic field. At very high magnetic field, $\triangle R = (R - R_s)$ is negative since the resistivity is further lowered by anisotropic magneto-resistance. The arrows in the circles indicate the direction of the magnetization in the Fe

layers. When the directions of magnetization in the two iron layers are antiparallel corresponding to zero applied field, the electrical resistance compared to R_s is the highest. As the magnetic field increases the magnetization in both the layers gradually Allin in the direction of the field and the resistance starts decreasing until the direction of magnetization in both the layers of iron Allin parallel to the direction of the applied field, in which case the change in resistance vanishes. Thus the resistance of the trilayer system attains its minimum value when its magnetization is ferromagnetic and maximum when it is antiferromagnetic. For an antiferromagnetic trilayer, this decrease in resistance could be as large as 20%. Considering the fact that the normal positive magnetoresistance in a metal is small, being only a small fraction of 1%, this negative magnetoresistance is giant. Hence, when discovered it was termed giant magnetoresistance (GMR). GMR is also observed in magnetic multilayers; where the decrease in resistance of the system on applying a magnetic field could be as large as 80% or more. The results of measured resistance as a function of the applied magnetic field for such multilayers are depicted in Fig. 7.

The figure shows the resistivity relative to its zero field value as a function of the magnetic field; for three different Fe/Cr magnetic multilayers. The notation $(Fe30\text{\AA}/Cr12\text{\AA})_{35}$ denotes that a multilayer structure is built by repeating 35 times the sandwich of $30\text{\AA}$ thick magnetic Fe layer and $12\text{\AA}$ thick nonmagnetic Cr layer, which are metallic magnetic and non-magnetic multilayers. One

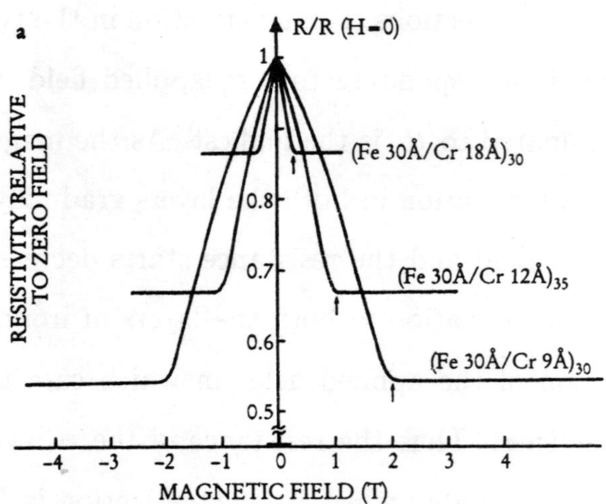

Figure 7: Percentage change of resistivity as a function of magnetic field for *Fe/Cr* multilayers (taken from P. Grünberg[2]).

can also make multilayers of magnetic metallic and insulating non-magnetic sandwiches. In the later case the electrical conduction will be due to tunneling of electrons. The arrows in the figure mark the switching field for each multilayer. The thinner the non-magnetic layer the more is the switching field and the larger is the decrease in resistance on applying the external magnetic field.

As discussed earlier the interlayer exchange coupling (IEC) between magnetic layers separated by a non-magnetic metallic layer is responsible for the antiferromagnetic alignment of the magnetization in the magnetic layers which in turn leads to such conducting behavior of the magnetic multilayers in presence of magnetic field. But the main mechanism for GMR is considered to be the spin-dependent scattering of electrons in the magnetized layers.

A schematic representation of how spin-dependent scattering

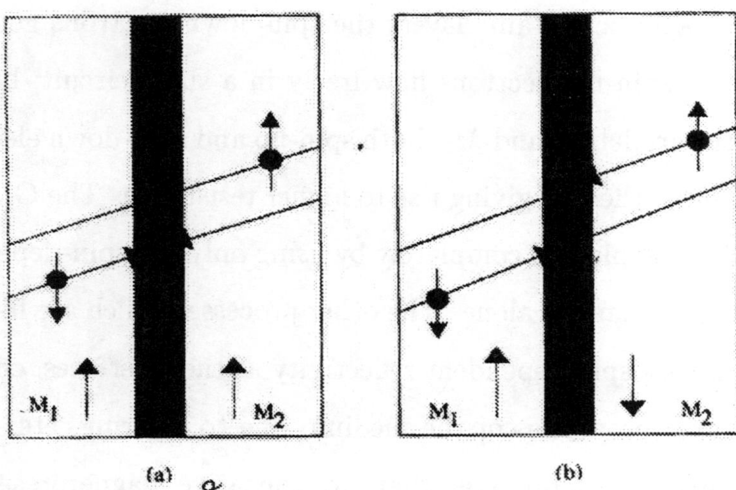

Figure 8: A schematic representation of the spin-dependent scattering mechanism, in magnetic multilayers, (taken from P. Grünberg[2]).

may lead to the variation of conductivity with the relative orientation of magnetizations in coupled layers, is shown in Fig. 8. The figure represents the cross section of a structure of two ferromagnetic films with magnetizations M_1 and M_2 in the directions indicated by arrows while these are separated by a non-magnetic metallic layer (the shaded region of Fig. 8(a) and Fig. 8(b) respectively). The net current is flowing from bottom to top while the conduction electrons (dots with an arrow denoting the spin orientations) drift randomly within and between the layers. For simplicity it is assumed that the electrons with their spins anti-parallel to the direction of local magnetization are only scattered. Under this condition, (a) with M_1 and M_2, the magnetizations aligned parallel in the two magnetic metallic layers the spin-down electrons get scattered while spin-up electrons flow freely in a short circuit; but (b) with anti-parallel M_1 and M_2, both spin-up and spin-down electrons are scattered, thereby giving rise to higher resistance. The GMR effect can't be explained completely by using only the spin-dependent scattering mechanism alone. The other processes which are likely to play a role are spin-dependent reflectivity at the interfaces, changes in electronic properties of the medium due to layering, etc. More recently it has been observed that large negative magnetoresistance is also exhibited by the manganite single crystals having the composition $La_{1-x}Sr_xMnO_3$. In these systems the observed decrease in resistance could be couple of 100% in the presence of an applied magnetic field. Hence such systems are said to exhibit Collosal Magnetoresistance (CMR). The electronic structure and the physical phenomena associated with the CMR systems [4] are much more complex compared to magnetic multilayers.

7   Half Metallic Magnets

While discussing the giant magnetoresistance in magnetic multilayers it was argued that electrons with different spin orientation propagate differently in these systems. Hence in principle the property of conduction by the spin polarized electrons can be utilized to make devices. So far electronic devices were always based on only one attribute of the electron namely its charge. Besides the "charge" the electron also carries a second attribute namely the spin; which can as well be used to make electronic devices. Lately there has been a lot of activity in this new field of spin electronics or shortly "spintronics". For the realization of such devices it is necessary to produce spin polarized electrons; which can of course be achieved by injecting electrons into a ferromagnetic material. But it is also possible that there may exist some materials in nature, in which electrons with only one spin polarization is responsible for the electrical conduction in those materials. Such materials are called "Half Metals"; and these have to be differentiated from normal metals and semimetals. In order to understand how such a thing can happen, let us first briefly review our understanding of metals.

According to band theory of solids, any material whose uppermost occupied band is partially filled with electrons so that the Fermi level (the highest occupied energy level in that band) lies within the band; then the material is a metal capable of carrying an electrical current. On the other hand if the highest occupied band is completely filled (usually designated as the valence band) and the next empty band (called the conduction band) are separated from each other by a forbidden gap, whose magnitude is of the order of an electron volt; the material is designated a semiconductor. If the magnitude of the band gap is of the order of tens of electron volts, then the material is an insulator, a bad conductor of electricity. Normally a fourth possibility exists, namely the filled valence band and the empty conduction band may have a small overlap or

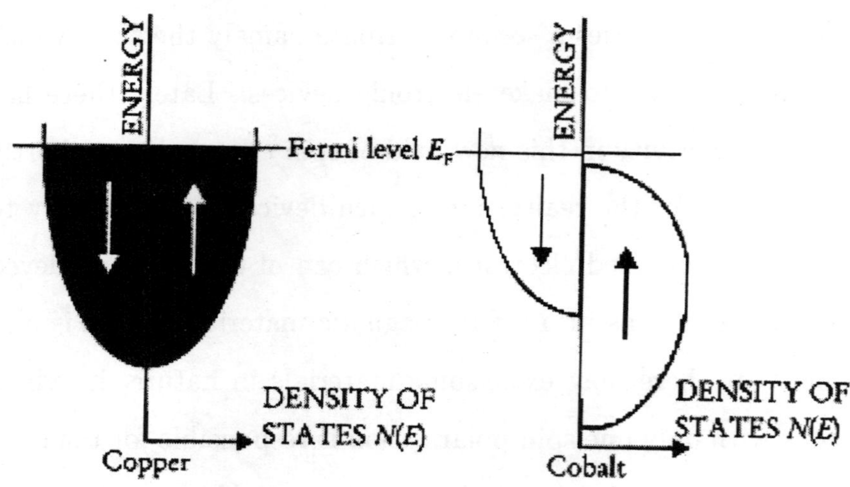

Figure 9: Density of states N(E) in copper and cobalt, represented schematically. Here the electron energy E is measured from the Fermi level E_F, the top of the filled states [5].

inverted band gap; such a material is then called a semimetal.

In metals as well as semimetals there are equal number of electrons with up and down spin and both of these contribute to the electrical conduction. In contrast in ferromagnetic metals electrons with one spin orientation is always more in number than the other, because of which the material acquires a magnetization. This difference between a normal metal and a ferromagnetic metal is schematically depicted in terms of the spin polarized density of states and the Fermi level in Fig. 9. In the later case the density of states corresponding to the two different spin directions of the electron are displaced in energy because of the splitting caused by the exchange interaction, which is responsible for the magnetization of the material. A more realistic picture of the spin polarized density of states and their occupation by electrons in the case of an elemental semiconductor like Silicon (Si), a ferromagnetic metal like iron (Fe) and a more complicated system like Chromium dioxide (CrO_2) is shown in Fig. 10. It can be seen from the figure that for Si, which is nonmagnetic, the spin up and spin down density of states are identical and the filled valence band is separated from the empty conduction band by a band gap in the middle of which lies the Fermi level. In the case of Fe which is a ferromagnetic metal, the density of states of the spin down electrons is shifted to higher energy as compared to that of the spin up electrons and the Fermi level lies within the band. As a result there are more spin up electrons compared to spin down electrons, which is responsible for the magnetization of iron. In contrast a schematic representation of the expected spin polarized density of states for chromium dioxide (CrO_2) is depicted in the lower panel of Fig. 9. In this case also the density of states of the spin down electrons of the conduction band is shifted to higher energy compared to that of the spin up electrons, but the Fermi level lies in the spin up band. The spin down band being separated by an energy gap from the Fermi level is unoccupied. Consequently in this material the electrical conduction is entirely due to the spin up electrons. Thus this material is expected to be a half metal; because it is an electrical conductor for spin up electrons and an insulator for spin down electrons. In the later case the Fermi level lies in the middle of the forbidden gap. This is purely a consequence of the crystalline structure of the material. Of course by definition any half metal will be a magnet, as it contains electrons only with one spin polarization. Since the half metal is magnetic, one of its constituents has to be an atom with a magnetic moment, like an element of transition metal or rare earth series in the periodic table.

The binary compound CrO_2 which is a half metal contains the transition metal atom Chromium (Cr). Besides this other solids which exhibits half metallicity and magnetism are ternary compounds in general and spinels in particular; which are minerals having the general formula AB_2O_4, for example Fe_3O_4, which can also be expressed as $Fe\,Fe_2\,O_4$; yet other examples of half metallic magnets are the Heusler's alloys, with general formula A_2MnB such as Co_2MnSi; and half Heusler's alloys, i.e. $AMnB$, e.g., $NiMnSb$. As mentioned earlier, the structure and bonding of the spinels, Heuslers etc. are just complex enough to encourage gaps in the density of states; which is also a requirement for half metallicity. For example half metallicity in the simplest binary compound CrO_2 occurs because the exchange splitting; i.e., the difference in energy between the up and down spin bands is greater than the occupied band width of the up electrons. As a consequence all valence electrons of Cr have up spin and none are down; CrO_2 is therefore fully spin polarized. In general however, the valence electrons in half metals need

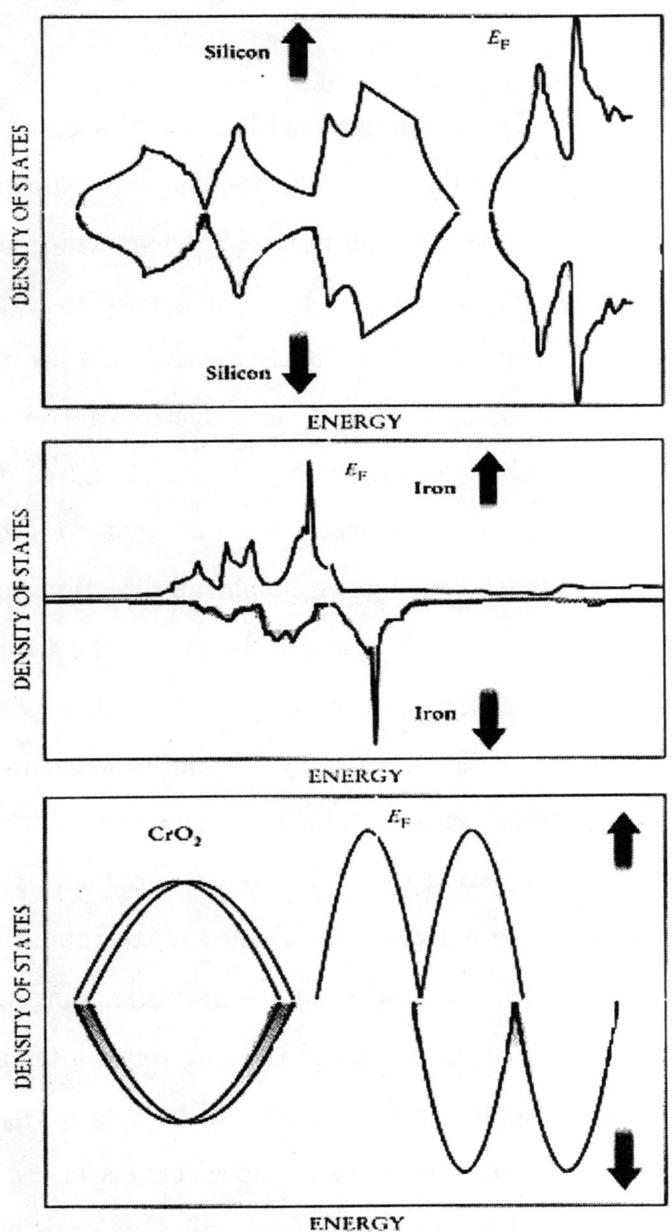

Figure 10: The density of states of silicon, iron and CrO_2[6]

not be entirely polarized but contain only an imbalance between up and down spin electrons.

Let us consider a more complicated half metallic oxide like Sr_2FeMoO_6. This compound has a double perovskite structure (perovskites have the general formula ABO_3) in which unit cells of the perovskites $SrFeO_3$ and $SrMoO_3$ alternate to form an ordered crystal. In this compound both Fe and Mo are magnetic ions; with different 3d-site energies and electronegativities. The electronic states in the solid therefore will consist of $Fe : d$, $Mo : d$ and $O : p$ electrons. The ionic states are Fe^{3+} with 5 electrons in the 3d-shell all of which are aligned in spin up direction because of exchange splitting; while Mo^{5+} has one electron in the 4d shell with spin down. Oxygen being in the ionic state O^{2-} requires 12 electrons out of which 3 are provided by Fe, 5 by Mo and rest four by the two Sr^{2+} ions. Thus the valence is satisfied making these compounds stoichiometric. Considering the fact that the Fe^{3+} and the Mo^{5+} magnetic moments are antiparallely aligned in neighbouring cells, with magnetic moments of $+5\mu_B$ and $-1\mu_B$ (μ_B being the Bohr magneton) respectively; the material is going to be a ferrimagnet. The net magnetic moment per unit cell will be $5\mu_B - 1\mu_b = 4\mu_B$. It should be noted that the five spin up electrons of Fe^{3+} fillup the five d-states resulting in a completely filled up spin band and hence do not contribute to electrical conduction of the material. Hence in Sr_2FeCoO_6 the metallicity comes from the spin down electrons, because only a fifth of the spin down states are occupied leading to metallic conduction. Usually in these materials the Crystal field splittings are smaller, while the interatomic couplings and hence the band widths are much larger; therefore the distinctions caused by the different magnetic orderings, such as ferromagnetic, antiferromagnetic and ferrimagnetic can effect the density of states drastically producing gaps at the appropriate energies resulting in the half metallic behaviour.

Yet another class of half metallic materials are diluted magnetic semiconductors (DMS) which are semiconductors containing a small concentration of magnetic impurities, so that it becomes a ferromagnetic semiconductor. Examples of these are $(Ga, Mn)As$ and $(Hg, Mn)Se$. These ferromagnetic semiconductors have two distinct band gaps one for each spin direction. When a small concentration of electrons or holes are doped or injected into these semiconductors, the carriers will only conduct if their spins are completely polarized. The diluted magnetic semiconductors are therefore the simplest half metals.

How does one experimentally distinguish a half metallic magnet from that of a magnetic metal? Do they have any distinguishing measurable properties? Indeed there are two such properties which differentiates a half metallic magnet from that of a magnetic metal, i.e., (i) these have quantized magnetic moments, and (ii) their magnetic susceptibility vanishes. Let us consider a half metal such that there exists an energy gap between a fully occupied down spin band and a partially occupied up spin band. To completely occupy the down spin band there must be as many down spin electrons per unit cell as there are energy levels. Hence the number of down spin electrons per unit cell $n \downarrow$ must be an integer. But for a stochiometric compound the total number of electrons per unit cell $n = n \uparrow + n \downarrow$ must be an integer. Hence the number of up-spin electrons $n \uparrow = n - n \downarrow$ also must be an integer. The magnetic moment per unit cell (m) bing $m = \mu_B(n \uparrow - n \downarrow)$, therefore must be an integer times the Bohr magneton (μ_B); which is quantized. By contrast both the up and down spin bands of a magnetic metal are partially filled, so neither the electrons with spin up nor spin down be an integer, nor is their difference. The magnetic moment of metallic Fe for example is $2.2\mu_B$ per atom.

In a conventional metal, arbitrarily small energy excitations are possible which can take an up spin electron to a down spin state or vice versa. But in a half metal arbitrarily small energy excitations are possible only when an up spin electron retains its spin direction; i.e. it can not go and occupy a down spin state. This is so because no low energy spin flip processes can occupy. Energy is required to flip a spin, i.e., to move an electron from an occupied spin-up state to an unoccupied spin down state or convert a spin down electron from the valence band below the gap to a spin up state above the Fermi level.

The magnetic susceptibility (χ) of a material measures how the magnetization changes on changing the applied magnetic field (H). In other words it is defined as the derivative of the magnetization

(M) with respect to H, i.e.

$$\chi = \frac{dM}{dH} = \frac{\Delta M}{\Delta H} = \frac{0}{\Delta H} = 0$$

A vanishing spin susceptibility means that the magnetization does not change as the applied field is varied. In a half metallic magnet, since all the electrons are already spin polarized, the magnetization does not change on applying a codirectional magnetic field. For example if all the electrons are in the spin up state, then there are either no electrons in the spin down state or it requires energy to bring a spin down electron from the valence band to the spin up state above the Fermi surface. Hence the spin susceptibility of a half metal vanishes.

8   Conclusion

In conclusion we summarize the new results associated with magnetic thin films and multilayers as discussed in this article, which are

1. As one approaches lower dimensionality by making the magnetic films thinner and thinner the direction of magnetization of the film remains in the plane until one reaches monolayer thickness (1.8 monolayers) when the direction of magnetization suddenly switches perpendicular to the plane of the film which is a surprising result.

2. When a magnetic thin film is deposited on an antiferromagnetic substrate, it always gets magnetized in a preferred direction as if the substrate provides a bias to the thin film. This phenomenon is denoted exchange bias, which awaits a proper understanding.

3. In a magnetic multilayer the magnetization is aligned either ferromagnetically or antiferromagnetically depending on the thickness of the sandwiched nonmagnetic metallic layer. Hence one can infer that the interlayer exchange coupling strength $J(R)$ oscillates as a function of the thickness of the monon-magnetic metallic layer in a magnetic multilayer system. The microscopic basis for such a coupling is as yet unknown.

4. Magnetic multilayers exhibit the negative giant magneto-resistance (GMR) effect whose magnitude depends on the thickness of the nonmagnetic metallic layers, as well as the number of magnetic layers.

5. A microscopic understanding of all these phenomena observed in a magnetic multilayer; in terms of the atomic spins in the magnetic layers, is yet to emerge.

6. There are materials in nature where all the conduction electrons are spin polarized. Hence unlike a normal metal where both the up and down spin electrons exist in equal number and both contribute to electrical conduction; in these materi-

als, electrons only with one spin orientation, are responsible for conduction. Hence these are termed "half metals" and by definition these are magnetic in nature. The half metallic magnets are characterized by the properties that, (i) their magnetic moments are quantized and (ii) their spin susceptibility vanishes.

The magnetic multilayers have already been used in technology to make devices. The property which finds application is the GMR effect. Using this magnetic read heads have already been built, yet another application of the GMR effect in magnetic multilayers is its possible use as a biosensor. An understanding of the production and transport of spin polarized electrons, in magnetic multilayers and contacts had led to the development of a new area in technology called spin electronics or "Spintronics". This is an active field of research at present. Yet another area where lots of excitement is generated is the discovery of photomagnetization in diluted magnetic semiconductors.

Acknowledgement: The transcript is based on talks delivered at several places. My sincere thanks goes to Mr. Santanu Maity for taking the notes of the talk delivered at Saha Institute of Nuclear Physics and for pain stakingly putting these notes into the form of a first draft. Without his efforts the transcript would never have seen the light of the day. I would also like to thank Professor R.K. Moitra for a critical reading of the manuscript. I would like to acknowledge Prof. G.S. Tripathi for getting me interested in the subject. I acknowledge the help rendered by Mr. D. Pradhan in preparing the manuscript.

References

[1] For a brief history of magnetism, see G.S. Tripathi, Ind. J. Phys. **77**A, (2003) 543.

[2] P. Grünberg, Physics Today, May (2001) 31.

[3] L.M. Falicov, Physics Today, October (1992) 46.

[4] Y. Tokura in "Colosal Magnetoresistive oxides Ed. Y. Tokura (Gordon and Breach Science Publishers, 2000) pp 1.

[5] G.A. Prinz, Physics Today, April (1995) 58.

[6] W.E. Pickett and J.S. Moodera, Physics today, May (2001) 39.

Physics of Solids, Nuclei and Particles
Editor: R. Sahu

Giant Magneto Impedance Phenomenon-Recent Results

S.K. Ghatak

Department of Physics and Meterology, Indian Institute of Technology
Kharagpur 721 302, India

INTRODUCTION

The influence of magnetism on charge transport is known for nearly a century. The most well-known effect of magnetic field is the force on charge carrier leading to Hall effect. The increase in carrier drift path in presence of magnetic field leads to magneto-resistance effect in pure metal. This magneto-resistance is positive, small and varies quadratically with magnetic field. In ferromagnetic metal and alloys a relatively large anisotropic magneto-resistance have been found and this additional resistance depends on relative angle between magnetization and current direction. The magneto-resistance can be enhanced in thin film geometry and large value is realized in sandwich structure of magnetic and non-magnetic metals and is referred to as giant magneto-resistance phenomenon[1]

Giant magneto-impedance (GMI) phenomenon refers to the observation where a large and sensitive change in impedance can be induced by reorienting the magnetization in soft magnetic material [2-5] The GMI can be looked as high frequency analogue of giant magneto-resistance. However, the former has higher sensitivity at low field compared to latter case. Besides, there are hysteresis and thermal stability problem in magneto-resistance. On the contrary GMI material exhibit nearly zero hysteresis when the field is swept and also has larger speed of response. In term of sensitivity GMI element can reach lower limit of SQUID. This large change in impedance has been observed in amorphous ferromagnetic materials with very low magnetic anisotropy, and this effect has created a growing interest among the researchers because of their promising technical application in recording heads and magnetic sensor [6,7].

GMI EFFECT

The magneto-impedance effect involves the change in impedance in ferromagnetic conductor excited by a.c. magnetic field and subjected to d, c. magnetic field. Consider a cylindrical conductor of length l and uniform cross-section carrying current $I = I_0 e^{-i\omega t}$ (Fig.-1)

The voltage drop V_0 across length l is related to impedance Z as:

$$Z = \frac{V_0}{I_0} = Z_r + iZ_i$$

Where Z_r and Z_i are resistance and reactance components respectively. The magneto-impedance (MI) is expressed in terms of ratio

$$\frac{\Delta Z}{Z} = \frac{Z(H) - Z(H_m)}{Z(H_m)}$$

Where H_m is maximum d.c. Field applied along length of the conductor The MI is weak in non-magnetic conductor but it exceeds few hundreds in soft ferromagnetic conductor when the maximum field H_m is of order of few Oe. This phenomenon is referred as giant magneto-impedance (GMI) effect. Depending upon relative orientation of exciting field h and bias field H, GMI effect can be classified as longitudinal (h & H are parallel) or transverse (h & H are perpendicular). The arrangement in Fig.1 corresponds to transverse case as exciting field is circumferential and H is along the

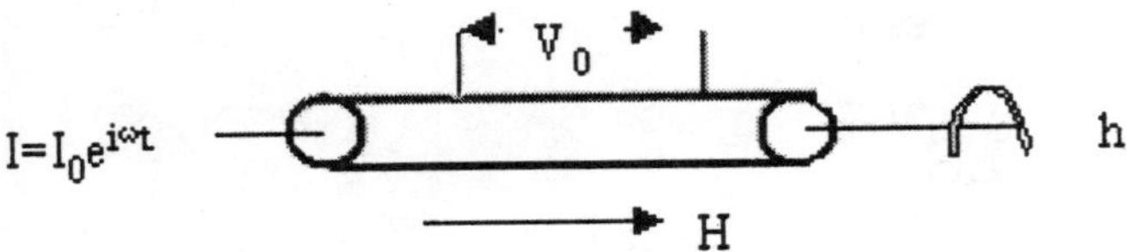

Figure 1: Schematic diagram for MI measurement

axis. The current density distribution across cross-section of conductor is non-uniform due to screening of effect. In ferromagnetic
metal conduction electrons and time varying magnetization cause
the screening of time dependent electric field. This induced magnetization depend on number of parameters some like saturation
magnetization, anisotropy field, relaxation time etc- intrinsic
to material. The domain structure in ferromagnetic state is governed by anisotropy field, which originates from combined effect
of spin-orbit coupling and ligand electric field, stress and geometry of sample. In structurally disordered materials -in particular
magneto-crystalline anisotropy nearly vanishes due to complete re-
moval of orbital degeneracy of ground sate of magnetic ion. In such
a system anisotropy is mainly magnetostrictive in nature and its
magnitude depends upon frozen stress due to fast quenching from
molten state. This anisotropy can be much reduced by appropriate
thermal treatment. Smaller anisotropy leads to large magnetic response. The response is also strongly influenced by the amplitude
and frequency of exciting field. Therefore the GMI ($\Delta Z/Z$) becomes function of saturation magnetization M, anisotropy filed H_k,

Stress σ, thermal history, frequency and amplitude of excitation of a.c. exciting field. Below representative results of GMI on these parameters are presented.

MATERIALS

GMI materials share property of magnetic softness. The soft magnetic materials are easy to magnetize and has large permeability with low coercive field. They possess narrow hysteresis loop with low core loss. Alloy of Fe-Si (3-4%) is a soft ferromagnetic materials - commonly used in motors, power transformer and generator. They are now available in the form of wire(typical diameter of the order of micron).ribbon ,magnetically coated metallic (non-magnetic) wire (tubes) thin film and multi- layers. Metallic glass prepared by rapid cooling are normally soft ferromagnet. The softness is due to negligible magneto-crystalline anisotropy in amorphous state. The attractive soft magnetic properties are exhibited by alloy of 3d transition metal (e.g. Fe,Co,Ni) and glass former (like Si,B,P etc) with concentration near eutectic. An amorphous alloy $(Co_{0.8}Fe_{0.06})_{72.5}s\ \dot{S}i_{12.5}B_{15}$ with magnetostriction $\lambda \sim -10^{-7}$ exhibit excellent soft magnetic behaviour. This is obtained by alloying FeSiB ($\lambda \sim 25x10^{-6})alloy, soft ferromagnet deposited on non - magnetic conductor, poly$ with CoSiB ($\lambda \sim -3x10^{-6}$). Other example of amorphous materials exhibiting GMI are given in table-1 [8]. The GMI also exhibited by crystalline alloy like permalloy, soft ferromagnet deposited on non-magnetic conductor ,polycrystalline manganite multi-layers and μ-metal sheet. They all share soft magnetic characteristics.

Two family of alloy termed as "Finemet" and "Nanoperm" are nanocrystalline with very good magnetic properties at high frequency and is comparable to some of best (relatively expensive) Co-based alloy. Average grain size (15 nm) in Finemet is lower than that (25nm) in Nanoperm.

Table 1: List of materials exhibiting GMI

Wire $d \sim 50 - 125$ μm	$a - (Co_{0.94}Fe_{0.06})_{72.5}Si_{12.5}B_{15}$
	$a - Co_{72.5}Si_{12Co.5}B_{15}$
	$a - Fe_{77.5}Si_{7.5}B_{15}$
	$Ni_{80}Mo_{4.2}Fe_{15.8}$ Permalloy
Microwire : $d \sim 20 - 30\mu m$	$a - Co_{68.5}Mn_{6.5}Si_{10}B_{15}$
Amorphous ribbon	$(Co_{1-x}Fe_x)_{75}Si_{15}B_{10}$ x $= 0.03 - 0.08$
	$Co_{67}Fe_{3.9}Si_{15.3}B_{13}$
	$Fe_{80-x}V_xB_{20}$
	$Fe_{73.5}Nb_3Cu_1Si_{13.5}B_{90}$ Finemet
Multilayer:	$Fe_{88}Zr_7Cu_1B_4$ Nanoperm
	$CoSiB/Cu, Ag, Ti/CoSiB$
	$FeNi/Cu/FeNi$
Deposited	$Co_6Fe_{20}Ni_{74}$ on CuB
	$Co_{90}P_{10}$ on Cu
	FeNi on CuBe
Manganite	$La_{1-x}Ca_xMnO_3$ $x = 0.67 - 0.4$
μ - metal stripe	$Ni - Fe - Cu - Cr$ alloy

                    Physics of solids, nuclei and particles

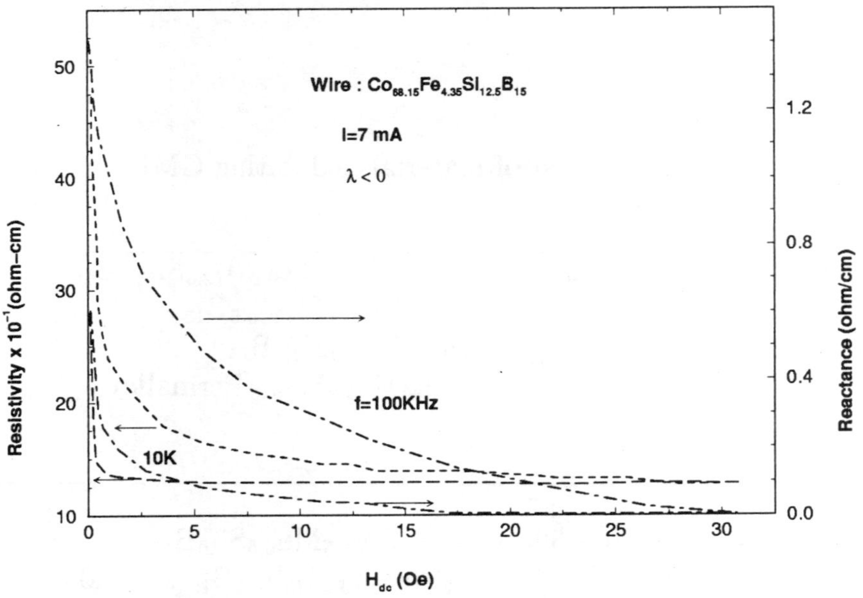

Figure 2: Field (H_{dc}) dependence of resistivity (solid) and reactance (dashed) of amorphous wire $(Co_{0.94}Fe_{0.06})_{72.5}Si_{12.5}B_{27.5}$ at T-300K for frequency f=10KHz and 100KHz and excitation current 7mA.

EXPERIMENTAL RESULTS:

1. **Field dependence** :The field dependence of the resistance and reactance of - amorphous wire $(Co_{0.94}Fe_{0.06})_{72.5}Si_{12.5}B_{15}$ is shown in Fig.2 [9]. The system is a soft ferromagnet with very small negative magnetostrictive coefficient. The exciting current of amplitude 7mA and frequency f=10KHz and 100KHz produces circumferential a.c. field and the d.c field is along the axis of the wire. Both the resistance (solid line) and reactance (dashed line) sharply decreases with H_{dc} and saturates at few Oe field. The maximum value of impedance at KHz region appears at zero field . However as frequency of a.c. field increases the position of maximum shifts to finite field that increases with frequency. The field where impedance nearly saturates depends sensitively on the magnetic softness

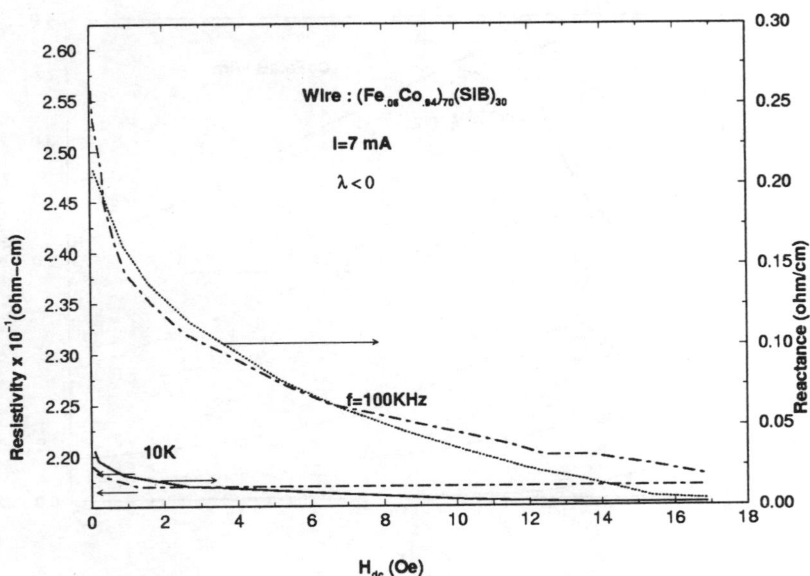

Figure 3: Field (Hdc) dependence of resistivity (dot-dashed) and reactance (dotted) of amorphous ribbon $(Fe_{.06}Co_{.94})_{70}(SiB)_{30}$ at T-300K for frequency f=10KHz and 100KHz and excitation current 7mA.

of the wire. Fig.3 depicts the similar results on resistance and reactance of amorphous ferromagnetic ribbon of nominal composition $Co_{66.01}Fe_{3.99}Si_{12}B_{18}$. This composition of Co-Fe alloy belongs to low magnetostrictive material family. As magnetic anisotropy in amorphous state is mainly due magnetostrictive in nature, the magnetic domains are oriented perpendicular to rolling direction of ribbon. The resistance tends to its d.c. value at a field greater than 1Oe for low frequency of excitation whereas larger field is needed for saturation for higher frequency.

2. **Amplitude & Frequency dependence** :Magnetic softness introduces non-linearity in response and is reflected in dependence of resistance and reactance on amplitude of excitation current (Fig.4). Both resistance and reactance at $H_{dc} = 0$

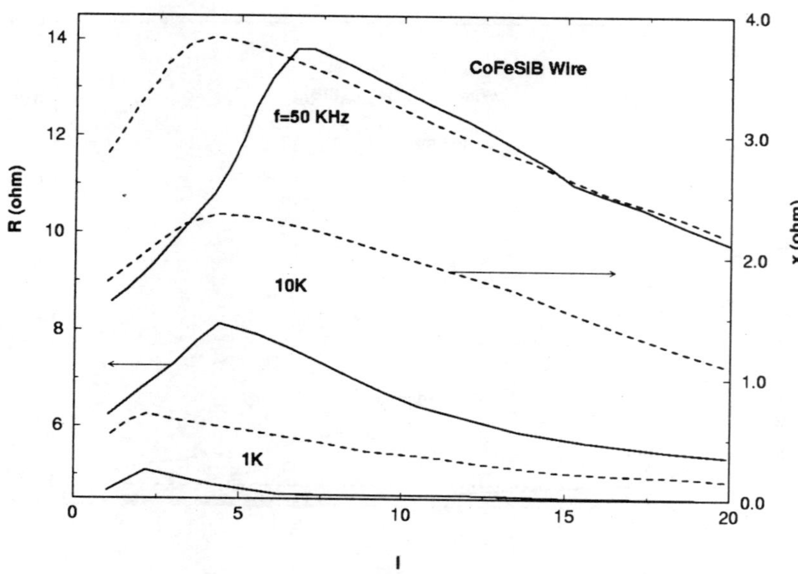

Figure 4: Variation of resistance and reactance of amorphous wire $(CoFe)_{72.5}(SiB)_{27.5}$ as a function of amplitude of excitation current at frequency f=100KHz.

show maximum value that again depends on frequency of excitation. At very small value of excitation magnetic field the Ohm law is obeyed. Above saturation value of Hdc this nonlinear response vanishes. The excitation current for maximum value is different for two components of impedance. The

impedance of amorphous ferromagnet at $H_{dc} = 0$ is an increasing function of frequency of excitation [2] (Fig.-5). Reactance exhibits stronger variation compared to resistance with frequency of excitation. However, in presence of large d.c. field resistance and inductance of the sample looses sensitivity of frequency dependence.

3. **Stress dependence** : The external stress can modify the domain pattern by altering the magnetic anisotropy and thereby

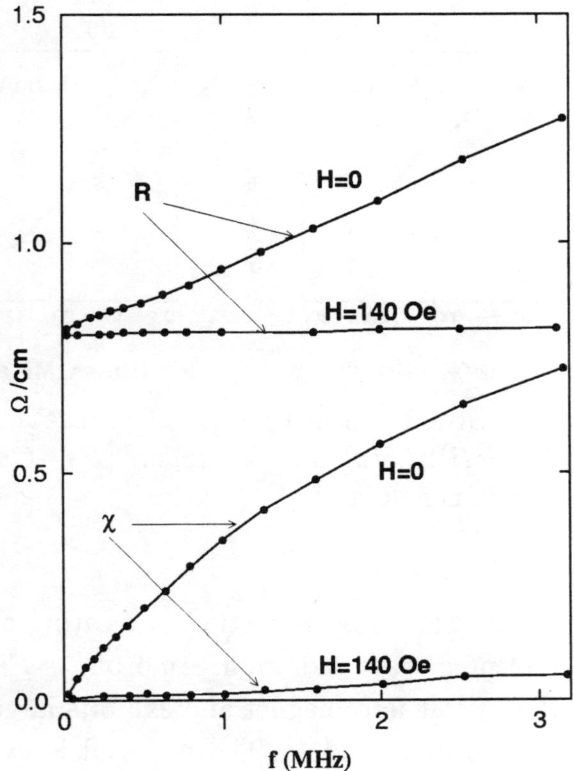

Figure 5: Frequency dependence of resistance R and reactance X per unit length of amorphous FeCoSiB wire for $H_{dc} = 0$ and 140 Oe. The driving current is 10mA rms

influences the impedance of these amorphous ferromagnet at $H_{dc} = 0$. Typical such influence is demonstrated in Fig.6 [10]. The longitudinal stress lowers the reactance at $H_{dc} = 0$ of $a - (Co_{0.94}Fe_{0.06})_{72.5}Si_{12.5}B_{15}$ wire (Fig.6a) and $Co_{66.01}Fe_{3.99}Si_{12}B_{18}$ ribbon (Fig.-6b). Note the difference in nature of variation of reactance with stress in amorphous wire and ribbon.

4. **Annealing dependence** :The domain pattern in amorphous state of soft ferromagnet can easily be changed by thermal treatment that facilitates stress relaxation. A simple treat-

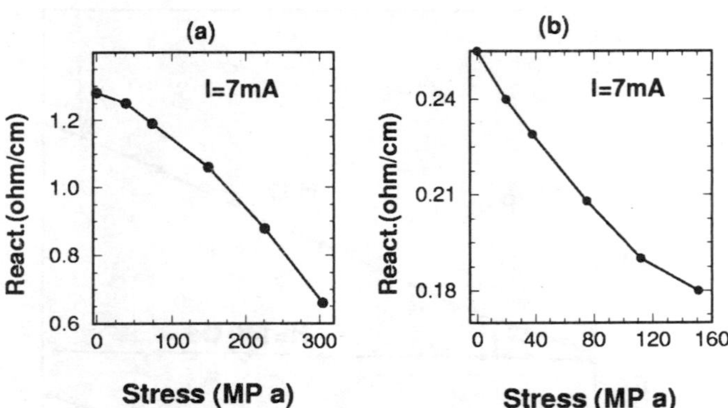

Figure 6: Stress sensitivity of reactance of a-wire $(CoFe)72.5(SiB)27.5(a)$ and a- ribbon $(Fe_{.06}Co_{.94})_{70}(SiB)_{30}$ (b) in absence of bias field and at 100KHz.

ment that causes stress reduction is treating the sample in presence of microwave radiation. Amorphous $Fe_{73.5}\,Nb_3\,Cu_1\,Si_{12.5}$ alloy is soft ferromagnet and exhibits large variation of magnetic impedance [11] . The material is exposed to microwave radiation by placing inside microwave woven for 10 minutes and taken out immediately. The measurement of MI is repeated in identical condition. A drastic change in behaviour in both resistance and reactance is found (Fig.7)[11]. Both quantities sharply fall to their saturation value at much lower field compared to as-quenched material. This sharp response to external field in annealed sample is associated with reduction of quenched stress with consequent reduction of magnetic anisotropy. The energy absorption by the material within the microwave oven is directly proportional to the out-of phase component (χ'') of magnetic response due to microwave magnetic field, frequency and field amplitude. As χ in ferromagnetic state can have appreciable magnitude ,the microwave exposure provides sufficient energy for the stress relaxation. More work is in progress to explore the efficacy of

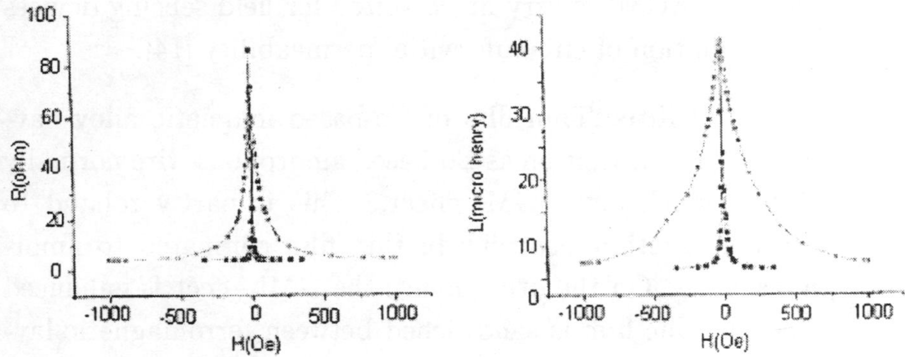

Figure 7: Field dependence of resistance and reactance of $a - Fe_{73.5}Nb_3Cu_1Si_{12.5}$ ribbon at frequency of 1MHz. Inner curve is the result after sample is exposed to microwave radiation for 10 min and outer one is that of as-quenched sample.

the microwave treatment for improvement of magnetic softness of the amorphous material. Preliminary data show formation of nono-crystalline structure for long exposure of radiation.

5. **Asymmetrical GMI**: A bias field (d.c or a.c) I in the direction of a.c. current introdusec asymmetry in the GMI with respect to sensing field.[12-13]. in twisted amorphous wire having helical magnetic anisotropy. Without bias field two symmetrical peak appears in MHz frequency range when sensing field is swept from positive to negative. The bias field reduces one peak and increases the other within higher sensitivity in an otherwise identical experimental condition.. By connecting two oppositely biased asymmetrical G/MI elements a linear voltage response can be obtained from larger region of sensing

field. The AGMI is very much suited for field sensing devices
and estimation of circumferential permeability [14].

6. **GMI in Films** :Thin film of Co-based magnetic alloy having similar composition as Co-based amorphous wire normally
exhibits much lower GMI effect. This is partly related to
higher magnetic anisotropy in thin film compared to amorphous state. On the other hand, the GMI effect is enhanced
when metallic film is sandwiched between ferromagnetic layers. In multi-layer film (like CoFeSiB/Cu/ CoFeSiB) large
GMI was observed [15-17]. Due to large resistance in magnetic
film the current through metallic film dominates and changes
magnetic structure of adjacent magnetic layers leading to enhanced GMI effect.

Applications

Many possible applications of GMI effect for sensitive and quick
response magnetic sensor are being explored. Areas of application include non-destructive testing, highly accurate rotary encoder
heads, medical electronics and auto-industry. With appropriate
choice of material properties that can be tailored for fast and large
impedance response sensor head can be designed. High sensitivity
to excitation current ,external stress and d.c magnetic field makes
these materials a good candidate for sensor application. Simplest
way to design field sensor is to use GMI element as inductor in
self-sustaining oscillator whose frequency of oscillation changes as
external magnetic varies [18].A low power advanced circuit using C-
MOS IC multivibrator is developed [19]. Using GMI element wide
variety of novel sensor including position , stress, current, biomedi-
cal and environment sensors have been developed [18,20,21].

THEORETICAL CONSIDERATIONS

Cylindrical Geometry : Wire

In presence of axial current a circumferential magnetic field is generated. For negative magnetostrictive material domain in ferromagnetic state is oriented perpendicular to axis of the wire (Fig- 8).

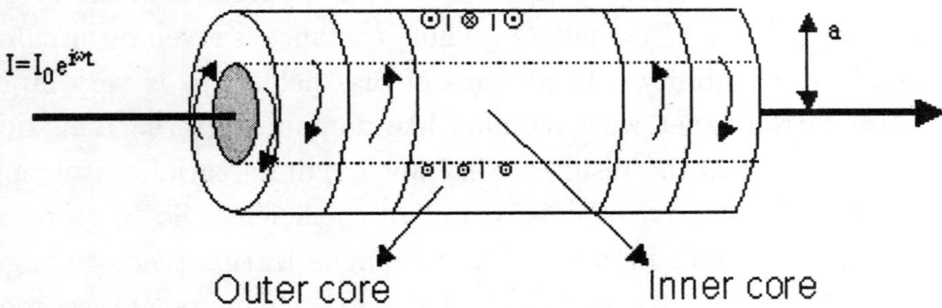

Figure 8: Geometrical configuration of amorphous wire. Magnetization in outer layer is circumferential with alternate right-handed and left-handed domain pattern. Domain within inner core is longitudinal.

Assuming linearity and isotropy ($m = \chi_t h$, $\chi_t = $ circumferential susceptibility)of magnetic state the impedance Z can be easily be obtained from Maxwell's equations and is given by [22]

$$Z = R_{dc}\frac{kaJ_0(ka)}{2J_1(Ka)}$$

with $k = (1 - i)/\delta_m$, $\delta_m = \frac{\delta_0}{\sqrt{\mu_t}}=$ effective skin depth that is reduced from $\delta_0 = [2/\omega\sigma\mu_0]^{1/2}$, skin depth in non-magnetic metal at frequency ω by transverse permeability $\mu_t = 1 + \chi_t$. J_0 and J_1 are the Bessel function of first kind and a is radius of the wire. The GMI behaviour can be divided into three regimes i) low frequency ii) medium frequency and iii) high frequency depending on frequency of exciting magnetic field. In low frequency regime the current is nearly uniform across cross-section and current produces circumferential magnetic field. This field h in turn induces mt and assuming homogeneity and isotropy , $b_t = \mu_t$ h. This regime is characterized by large skin depth so that $ka \ll 1$ and impedance $Z \approx R_{dc} + i\omega L_i$ where $L_i = \emptyset\mu_t l/2$ is internal inductance of the wire . For length l of wire Li = mt l/2 . Thus reactance is given by circumferential permeability . In absence of bias field H μ_t is very large for soft ferromagnet with bamboo like domain pattern. This domain structure is the result of very low circumferential anisotropy due to small negative magnetostrictive coefficient. So large reactance follows. With increase of H the magnetization tends to align along axis of the wire and this reduces m_t . Thus reactance goes down to small value when μ_t becomes small for complete alignment of magnetization at large H. This is referred to Giant magneto-impedance effect [23]. In medium frequency regime of few MHz the skin depth becomes comparable to /less than the radius of the wire. For $\delta \ll a, Z/R_{dc} \approx [1 + i](\delta_0/a\sqrt{\mu_t})$ so that both resistive and reactive part are equally weighted by transverse permeability. The field -driven impedance change is then due to corresponding change in μ_t . As μ_t is drastically reduced in presence of H bias field, large reduction of impedance is therefore related to increase in effective penetration of electric field in ferromagnetic conductor. This indicates that Z could reach few hundred time of R_{dc} and monotonically decrease from its maximum value at H=0 to minimum one at high H. This is in contrast to experimental finding where maximum appears at $H \neq 0$ and frequency dependent. We note that classical skin-depth formulation does not account the presence of domain structure in system. At high frequency domain dynamics plays important role in response to ac field. The associated domain dynamics introduced a phase lag in induced moment and thereby the transverse permeability becomes complex $\mu_t = \mu_{ti} + i\mu_{to}$ where μ_{ti} and μ_{to} are respectively in-phase and out-of-phase components. Assuming a single relaxation time $\tau, \mu_{ti} = \mu_t(0)/[1 + \omega^2\tau^2] and \mu_{to} = \mu_t(0)\omega\tau/[1 + \omega^2\tau^2]$ where $\mu_t(0)$ is its d.c. value. Incorporating this into Z , two-peak structure of Z with H is obtained and the peak shifts to higher value of H as frequency of excitation increases[24]. Therefore the understanding of GMI phenomenon is intimately related to evolution of magnetic permeability. In ferromagnetic material owing to involved relationship between magnetization and field the permeability is tensor which not only depends on frequency and amplitude of excitation but also depends on anisotropy. A good theoretical treatment of GMI involves to find expression of transverse permeability incorporating a particular domain structure of the substance . In general the response to a.c magnetic field comes from domain wall motion and domain rotation. At relatively low frequency both processes contribute to permeability. However, the eddy current damping at higher frequency reduces contribution due to wall motion and rotation process dominate magnetization process. The anisotropy in soft ferromagnetic material can be modified by external parameter like external stress, annealing etc., and this in turn changes the magnetization process and alters the GMI . Thermal treatment can induces anisotropy such that easy axis of magnetization is along axis of the wire. In this case MI decreases monotonically and exhibits single peak.At higher frequencies magnetization is totally dominated by rotation process dynamic characteristic becomes more important. In such situation more complete treatment based on simultaneous solutions of Maxwell equations and Landau-Lifshitz equation for magnetization. The situation is akin to ferromagnetic resonance [25]. At higher frequency > 100MHz the results obtained from the above procedure show that the exchange effect in GMI response becomes important [26-28].

Rectangular Geometry : Ribbon & Film

The general configuration for MI measurement for ribbon or film is shown in Fig. 9.There is a similarity between ribbon and wire with circumferential anisotropy as equilibrium magnetization is transverse to excitation magnetic field. The probing current flows through ribbon of thickness $d \ll$ compared to other two dimensions.

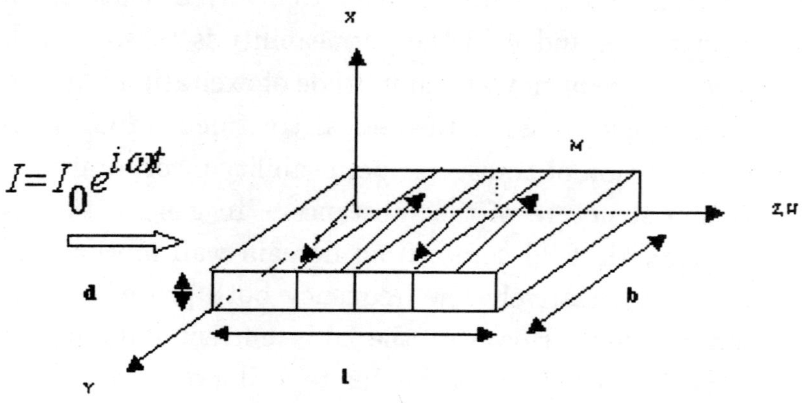

Figure 9: Geometrical configuration of amorphous ribbon with magnetization in the plane and directed nearly perpendicular to long axis of ribbon.

The transverse magnetization induced
I∥ z , h ∥ y & for in-plane easy axis M ∥ y
$h(x) \sim h_{y0} e^{jkx}$

Impedance $Z = R_{dc}(kd)\mathrm{Coth}(kd)$ with $k = (1 - i)/\delta_m, \delta_m = \frac{\delta_0}{\sqrt{\mu_t}}$ and transverse permeability $\mu_t = 1 + \chi_t$.

Conclusions

The giant magneto-impedance phenomenon appears to be classical in origin related to magnetic softness of the materials. Good examples are Co- based amorphous ferromagnetic alloy in shape of ribbon ,wire or microwire. The magnetic anisotropy energy in amorphous state is mainly magnetostrictive in nature ,and can be reduced to a small value either by decreasing magnetostriction coefficient or residual stress. The former can be varied by changing alloy composition and latter by different thermal treatment. Large effort are being given to optimize the magnetic softness and GMI response. Although qualitative understanding has been achieved based on classical electromagnetism the predicated response differs quantitatively. Moreover, as the response is highly nonlinear any theoretical explanation should consider incorporate the nonlinearity in treatment of MI. In field of application it is way ahead to understanding of the phenomenon. The advantage in GMI is the large decrease in impedance in very small field compared to that in GMR. Apart from application in sensor technology the GMI is considered as an additional tool to study the domain dynamics ,circumferential perameability and magnetostriction of soft ferromagnetic materials.

References

1. M.N.Baibich, J.M.Broto, A.Fert, F.N.van Dau and F.Fetroff, Phys. Rev. Lett. 61, 2472 (1988)

2. R.S.Beach, A.E.Berkowitz, Appl. Phys. Lett. 64, 3652 (1994)

3. R.S.Beach, A.E.Berkowitz, J. Appl. Phys. 76, 6209 (1994)

4. L.V.Panina, K.Mohri, Appl. Phys. Lett. 65, 1189 (1994)

5. F.L.A.Machado, C.S.Masrtins, S.M.Rezende,Phys. Rev. B 51, 3926 (1995)

6. M.Knobel, K.R.Pirota, J. Magn. Magn. Mater. 242, 33 (2003)

7. A.E.Mahdi L.Panina and D.Mapps , Sensors and Actuators A 105 ,271 (2003)

8. C.Tannous, J.Gieraltowski, arXiv:physics/0208035 Aug.(2002)

9. S.K.Ghatak , unpublished

10. B.Roy, S.K.Ghatak, J.Alloy and Compounds 326, 198 (2001)

11. B.Kabiraj , S.K.Ghatak, to be published

12. T.Kitoh, K.Mohri,T.Uchiyama, IEEE. Trans. Magn. 31, 3137, (1995)

13. L.V.Panina, D.P.Makhnovskiy, K.Mohri, J. Appl. Phys. 85, 5444 (1999)

14. C. Gomes-Polo, M.Vazquez, M.Knobel, Appl. Phys. Lett. 78, 246 (2001)

15. S.Q..Xiao, Y.H. Liu,Y.Y. Dai, L. Zhang,S.X. Zhou, G.D. Liu, J. Appl. Phys.,85, 4127 (1999)

16. Y.Zhou, J.Yu, X.Zhao, B.Cai, IEEE Trans. Magm. 36, 2960 (200),

17. A.E.Mahdi ,L.Panina, D.Mapps, Sensors and Actuators A 105, 271 (2003)

18. K.Mohri, T.Uchiyama, L.V.Panina, Sensors and Actuators A59, 1 (1997)

19. T. Kano, K. Mohri, T. Yagi, T. Uchiyama, L.P. Shen, IEEE, Trans. Magn.33, 3358 (1997)

20. M.Vazquez.M.Knobel, M..Sanchez, R.Valenzuela, A.Zukhov, Sensors and Actuators A59, 20 (1997)

21. W.S. Cho, H. Lee, C.O. Kim, Thin solid Films, 375, 51, (2000)

22. L. Landau, E.M. Lifshitz, Electrodynamics of Continuous Media, Pergamon New York, 1975

23. K. Mandal, S.K. Ghatak, Phys. Rev. B 47, 14233 (1993)

24. S.K.Ghatak -unpublished

25. Akhhiezer et al in Spin waves ,Wiley, New York 1968 Ed: S. Doniach

26. A.Yelon et al Appl. Phys. Lett. 69, 3084 (1996)

27. L.Kraus, J. Magn. Magn. Mater, 195, 764 (1999)

28. D.Manard et al ,J. Apl. Phys 88, 379 (2000)

Physics of Solids, Nuclei and Particles
Editor: R. Sahu
Copyright © 2006, Narosa Publishing House, New Delhi, India

Transport Through Quantum Dots

Sumathi Rao

Harish-Chandra Research Institute, Chhatnag Road, Jhusi, Allahabad 211 019, India

Abstract

We give an introduction to quantum dots and discuss why they
are interesting to study, explaining both theoretical issues and applications. Then we discuss some recent experiments on coupled
quantum dots. We discuss different ways to understand it and then
propose a model for it and discuss the results. We conclude with
ideas of further extensions, including the idea of quantum charge
pumping through multiple quantum dots.

INTRODUCTION

- What are quantum dots?[1]

---

[1]Talk at National Seminar on 'Recent Advances in Physics', 19-20 September
2003, Gopalpur-on-sea, organised by Berhampur University and Institute of
Physics, Bhubaneswar.

A quantum dot is an island containing a discrete number of electrons. The number of electrons on the dot can be changed by adjusting the electric field in the neighbourhood of the dot by applying a voltage to nearby metal gate.

As a circuit element, it is depicted below.

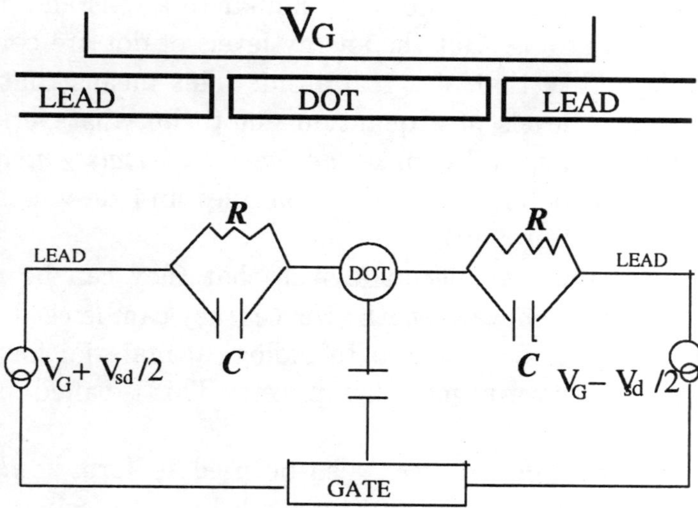

Single dot device and equivalent circuit

The typical dimensions of the dots range from $\sim$ 1-100 nanometer $\sim 1 - 100 \times 10^{-9}$ m. And typically they contain few hundred electrons. One can think of them as a constrained nanometer sized region of semiconductor material. One can have 1D dots, which are just lengths of one-dimensional wires between potential barriers or constrictions. Similarly, 2D dots are constrained areas which could be circular and 3D dots are constrained regions which could be spherical.

The main point is that constraints lead to quantisation of energy levels, as one knows from solving simple quantum mechanical problems like the particle in a box. So the energy levels of a quantum dot are discrete. And as one knows from usual quantum mechanics, the smaller the dots, the more separated are the energy levels. Closed quantum dots are isolated and have infinite hard wall potential barriers, whereas open quantum dots are coupled to leads through tunneling barriers, which are finite. Hence, electrons can leak out to the leads.

● Why study quantum dots?

There is a an enormous potential for applications of quantum dots. The energy levels of a quantum dot are analogous to energy levels of atoms. Therefore dots are similar to atoms and are often called artificial atoms. But the energy levels of dot are completely controllable, unlike that of a real atom. This means that we can tune the energy levels of a quantum dot to be whatever we like. That way, we can actually make *'designer materials'* - *i.e.*, we use artificial atoms to build artificial molecules and design materials with any required properties.

Artificial atoms can be made such that they can be made to fluoresce at any given wavelength. Hence, they can be used as nano-sensors. They can be attached to biological material for medical imaging, drug discovery, gene therapy, etc. This is called *'biological tagging'*.

Grids of quantum dots can also be used to form *'cellular automata'*. Research is underway to make logic gates using quantum dots and transmit signals and see if quantum dots can be used to build quantum computers.

THEORETICAL ISSUES IN QUANTUM DOTS

One of the most important consequences of the strong confinement of electrons is that it makes many-particle effects prominent. Also, since the dimensionality is low, the many-particle effects tend to be strongly correlated. In the last few years, for a large fraction of condensed matter theorists who work on strongly correlated electron systems, quantum dots and quantum wires have become important simplified 'lab' systems where many of the models and methods involving strong correlations can be explicitly tested.

• Coulomb blockade

A simple straightforward consequence of the Coulomb repulsion between electrons is a phenomenon called *'Coulomb blockade'*. The idea, at the semi-classical level, is that quantum dots, because of their small sizes, have small capacitances. Hence, the capacitive energy that is needed to overcome the Coulomb repulsion offered by the dot and add another electron to the system $e^2/2C$ (C is the capacitance of the dot) is quite large. (For macroscopic large capacitance conductors, this energy is negligible). However, by tuning the energy levels of the dot by a gate voltage, it is possible to shift the levels, so that one level is degenerate with the Fermi energy of the electron that we wish to add. At this energy, the electron does not have to overcome any energy. This is called the 'lifting of the Coulomb blockade'. This happens periodically as a function of the gate voltage. Hence, as we tune energy levels of the dot by a gate voltage as shown below, we see peaks in the conductance.

The total electrostatic energy of the system is given by $U(n) = \frac{e^2 n^2}{2C} - \alpha(e\,n)\,V_g$. The Coulomb Blockade is lifted (by changing gate voltage) when $U(n) = U(n+1)$. *i.e.*, whenever two charge states of a dot are degenerate there exists peak in the transmission. Hence, current turns on and off each time one electron is added to the dot. This has been experimentally seen[?] and is extremely important in applications as single electron transistors.

The well-known Kondo effect is the unusual phenomenon where resistance increases as temperature is cooled below a certain temperature called the Kondo temperature[3]. The explanation for this phenomenon comes from the fact that the spin of the impurity couples to the spin of the conduction electrons. This coupling becomes stronger as temperature decreases, thereby implying more impedance to the flow of current and hence increased resistance. The important point about this effect is that it is a many body effect between the conduction electrons and the localised electron.

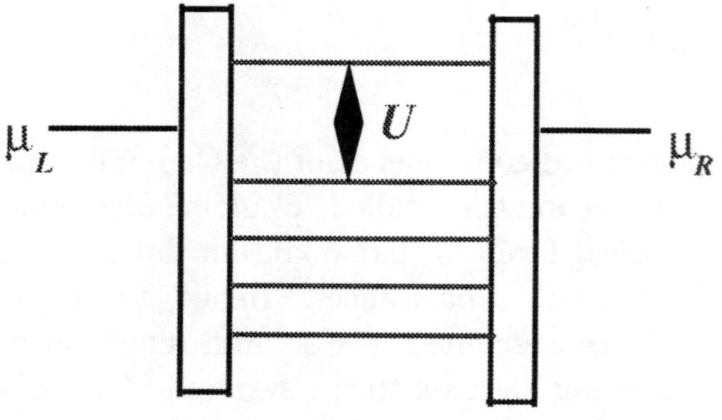

μ_L
U
μ_R

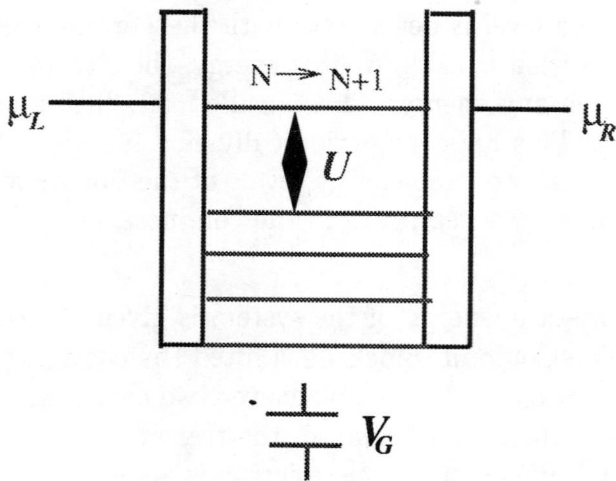

μ_L
N $\rightarrow$ N+1
U
μ_R
V_G

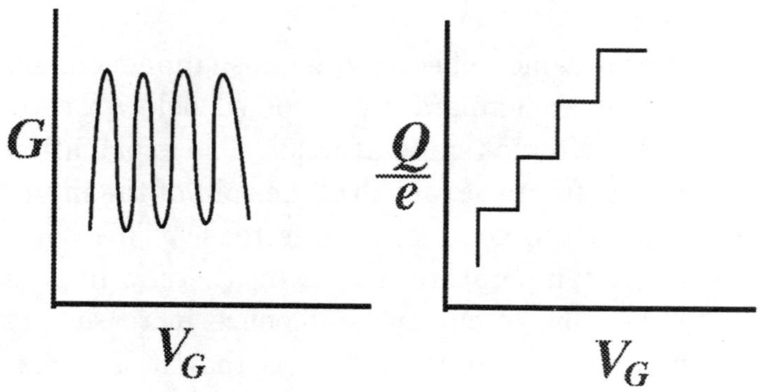

G
V_G
$\dfrac{Q}{e}$
V_G

• Kondo effect

A similar effect can be found in quantum dots. As has been already discussed, as we change the gate voltage by U, which is the Coulomb blockade energy and add electrons to the dot, the conductance curve forms a series of peaks. But below 'T_K, the Kondo temperature', peaks form pairs. The reason for this is that between paired peaks, the number of electrons on the dot is odd. Because of the many-body effect between the 'lead electrons' (analogous to the conduction electrons) and the electron on the dot (analogous) to spin, a bound state is formed that leads to enhanced conductance (instead of conductance going to zero). This is the manifestation of the Kondo effect in quantum dots.

Another way of saying it is to say that the unpaired electron forms a singlet with the electrons at the Fermi level in the leads, leading to an enhanced density of states at the Fermi level in the leads. This is the reason for the enhanced conductance. This has been experimentally seen in [4] and the description below is for the figures in that reference.

Normally the peaks become narrower as one decreases the temperature. (This can be seen in Fig. in [4] above 800 mK). But below 800 mK, the peaks broaden as one decreases the temperature. Here 800 mK is the approximate Kondo temperature T_K. Raising the temperature destroys the Kondo singlet and hence, above T_K, one gets the expected Coulomb blockade peaks and the conductance decreases between the peaks as expected. Below T_K, formation of the Kondo singlet enhances the conductance. From an experimental and theoretical standpoint, what is important about the Kondo effect in dots, is that it is a fully controllable Kondo effect. So one can actually measure details and compare it with theoretical predictions. This explains why quantum dots are of interest to theorists as well.

ISSUES IN COUPLED QUANTUM DOTS

We now present some recent experimental results on coupled quantum dots[5].

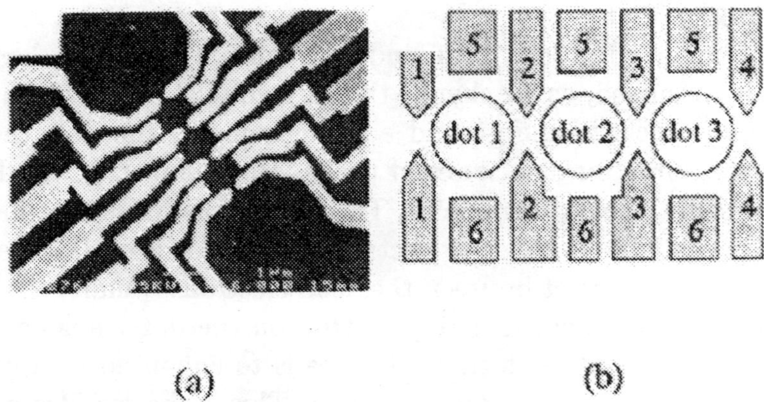

(a)   (b)

The relevant experimental details are given below.

$$\begin{aligned}
\mathrm{A_{dot}} &= 0.5 \text{ X } 0.8 \ \mu\mathrm{m}^2 \\
\mathrm{n_s} &= 3.7 \text{ X } 10^{11} \text{ cm}^{-2} \\
\mathrm{N_{electron}} &\leq 1500 \text{ per dot} \\
\text{Mobility, } \mu \text{ at } 10^0\mathrm{K} &= 5 \text{ X } 10^5 \text{cm}^2/\mathrm{Vs} \\
\text{Phase Coherence Length} &> 20 \ \mu\mathrm{m} \text{ for } \mathrm{T} < 1^0\mathrm{K}
\end{aligned}$$

The measurements were at temperatures as low as 100 mK.

$$\Delta\mathrm{E} \simeq 50 \ \mu\mathrm{eV} \qquad \mathrm{k_B T} \simeq 7 \ \mu\mathrm{eV}$$

$$\mathrm{U} \simeq \frac{\mathrm{e}^2}{\mathrm{C_\Sigma}} \simeq 400 \ \mu\mathrm{eV}$$

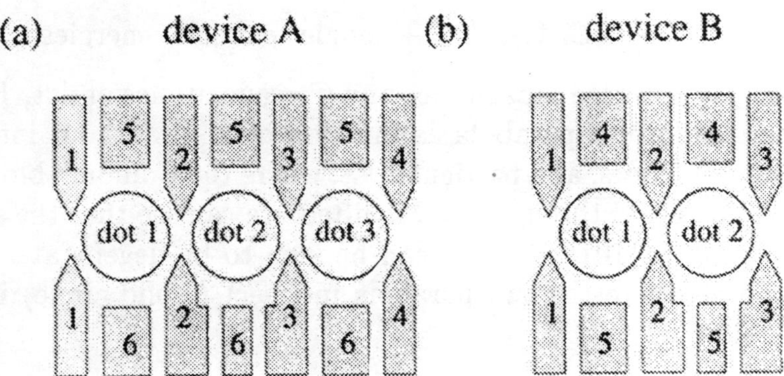

FIG. 1. Schematic diagrams of (a) triple dot (device A) and (b) double dot (device B). Dots are formed in a GaAs/Al$_x$Ga$_{1-x}$As heterostructure, and have a lithographic area 0.5×0.8 μm^2.

Quantum Dots in Series :

In terms of an equivalent circuit, coupled quantum dots in series can be depicted as below:

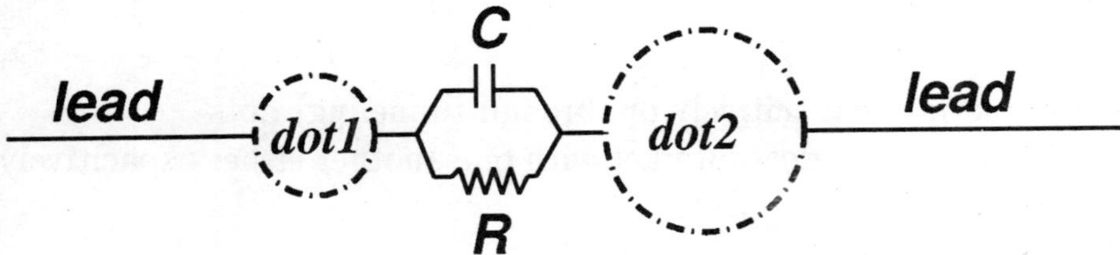

♣ Non-interacting dots :

For two non-interacting dots, the energy for the two dot system is given by

$$U(n_1, n_2) = \frac{e^2 n_1^2}{2C_1} + \frac{e^2 n_2^2}{2C_2} - \alpha_1(en_1)V_{g_1} - \alpha_2(en_2)V_{g_2}$$

where n_1 and n_2 are the numbers of electrons in each dot, C_i are the capacitances of the dots and V_{g_i} are the gate voltages on each of the dots. The Coulomb blockade in each dot is lifted by an independent gate voltage.

The single dot energies can also be written as

$$U = (ne - C_G V_G)^2 / 2C + \text{ single particle energies}$$

where C_G = dot-gate capacitance and C=capacitance of dot. Hence the energies form a parabola as function of V_G. Without interdot coupling, for 2 dots, at a particular V_G, where the Coulomb blockade through dot 1 and through dot 2 is lifted, we can see that the states $(n_1, n_2) = (00), (10), (01), (11)$ can be seen to be degenerate. This is the point where the two parabola intersect (solid circle) in the figure below.

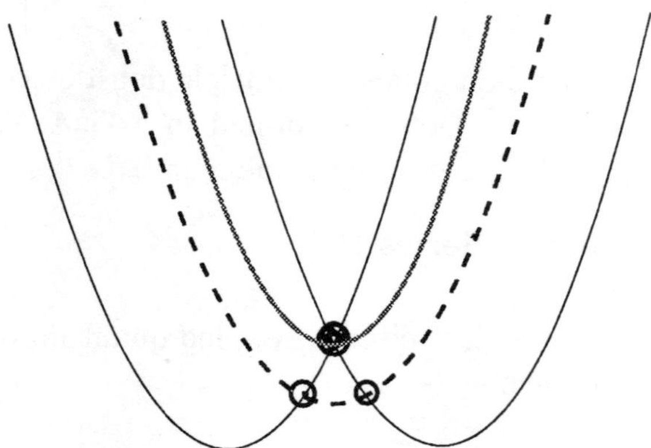

♣ Interacting (capacitively or through tunneling) dots

When the two dots interact with one another either capacitively or through tunneling, then

$$U(n_1, n_2) = \frac{e^2 n_1^2}{2C_1} + \frac{e^2 n_2^2}{2C_2} - \alpha_1(en_1)V_{g_1} - \alpha_2(en_2)V_{g_2} + E_{12}n_1 n_2$$

In this case, the Coulomb blockade through the two dots is coupled. The result of this is that it is not possible to simultaneously lift both the Coulomb blockades and make 4 states degenerate. Instead, the interdot coupling breaks the degeneracy between the parabolae and shifts one of them down. So at the hollow circles in Fig., only 3 states are degenerate. Hence, the conductance peaks split and the shift is proportional to Δ, which is the amount by which energy is lowered.

But quantitatively, numerical simulations of classical charging model need unrealistically large interdot capacitances to get the values of the splittings observed experimentally. Also, these models completely neglect tunneling, which is only correct if the tunneling rate is negligible. But in the experiments, the tunneling rate is actually $\sim e^2/h$.

The aim of our work was to give a quantitative theory for the conductance peak splitting.

• Our model and results

We model the dots as one dimensional dots[6]. The justification for this comes from Ref.[7], who showed that for sufficiently large dots, as long as the width of constriction only allowed a single transverse state below the Fermi level, the electron gas inside the dot behaves essentially one-dimensional.

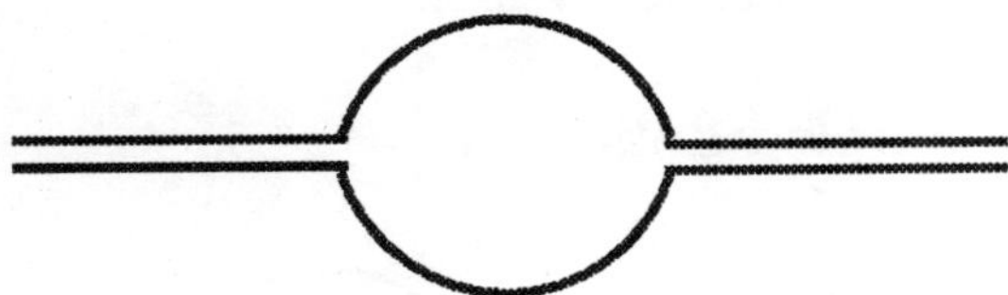

In fact, in [8], it was also argued that electrons in these dots form Luttinger liquids. The idea they used was that electrons inside circular dots form shells. Hence, if we only look at the electrons in the outermost shell, essentially one of the quantum numbers is frozen, and they behave like one- dimensional electrons. In fact, they claimed that their numerical calculations showed evidence that the outermost shell shows Luttinger liquid behaviour.

With this motivation, we model quantum dots as short lengths of Luttinger wire with barriers separating them from the Fermi liquid leads.

Luttinger liquids

The idea is that in one-dimension, the generic theory, for interacting electrons is Luttinger liquids, and there is no regime where Fermi liquid theory is applicable[9]. The steps that one takes to show Luttinger liquid behaviour is as follows. We start with a free fermion dispersion and linearise around the Fermi points to rewrite the the theory in terms of the Dirac fermions (linear dispersion). These fermions can be then rewritten in terms of the density perturbations which are bosonic(free massless bosons); in fact, a rigorous derivation shows that all excitations in terms of fermions can be matched by excitations in terms of bosons. We can also then show that inter-electron interactions (four-fermion terms) continue to be quadratic in the bosonic language, as long as we ignore backscattering. This makes it extremely easy to study interacting theories as free bosonic theories. The Lagrangian for the bosonic theory can be written as

$$\mathcal{L}(\phi; K, v) = (\frac{1}{2}Kv)\,(\partial_t\phi)^2 + (\frac{v}{2K})\,(\partial_x\phi(x))^2$$

The interaction parameter K varies from 0 to 1 $K = 0$ corresponds to strongly interacting fermions and $K = 1$ corresponds to free fermions v is the quasi-particle velocity.

Our model of coupled dots

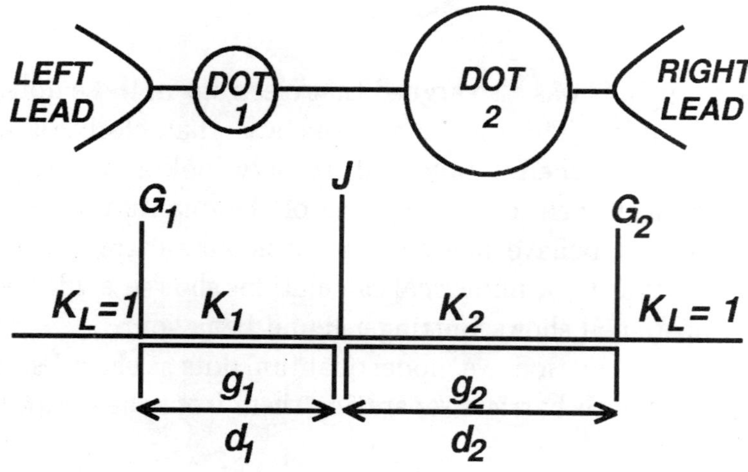

For our model with coupled dots, the Luttinger liquid parameters are K_1 and K_2, the barrier heights are given as G_1, G_2 and J and the gate voltages are g_1 and g_2.

The action for our model is given by

$$\mathcal{S} \;=\; \int d\tau \left[\, \mathcal{S}_{\text{leads}} \;+\; \mathcal{S}_{\text{dots}} \;+\; \mathcal{S}_{\text{gates}} \,\right]$$

$$\mathcal{S}_{\text{leads}} \;=\; (\int_{-\infty}^{-d_1} + \int_{d_2}^{\infty}) dx \mathcal{L}(\phi; K_L = 1, v_F)$$

$$\mathcal{S}_{\text{dots}} \;=\; \int_{-d_1}^{0} dx \mathcal{L}(\phi; K_1, v_1) + \int_{0}^{d_2} dx \mathcal{L}(\phi; K_2, v_2)$$
$$+ \;\; V + J\cos(2\sqrt{\pi}\chi)$$

$$\mathcal{S}_{\text{gates}} \;=\; g_1 \int_{-d_1}^{0} \rho(x,\tau) + g_2 \int_{0}^{d_2} \rho(x,\tau)$$

$$\text{with} \quad \mathcal{L}(\phi; K, v) \;=\; (\tfrac{1}{2}Kv)\,(\partial_t \phi)^2 + (\frac{v}{2K})\,(\partial_x \phi(x))^2$$
$$\text{and} \quad V \;=\; \sum_{i=1}^{2} G_i \cos(2\sqrt{\pi}\phi_i + (-1)^i 2 k_F d_i)\;.$$

G_1, J and G_2 are the barrier strengths at $x = -d_1, 0, d_2$ respectively and ϕ_1, χ and ϕ_2 are the quantum mechanical field variables at $x = -d_1, 0$ and $x = d_2$ respectively.

The Coulomb blockade through both dots are coupled because of tunnel coupling. (Electrons can tunnel through the δ-function barrier whose strength is J.)

First, we consider the case where $J \to \infty$, which, in turn, implies two independent dots with their independent Coulomb blockades(CB). We can plot the CB maxima as a function ofg_1, g_2 and we get a square grid, since the maxima for each dot is periodic in the gate voltage that tunes its maxima.

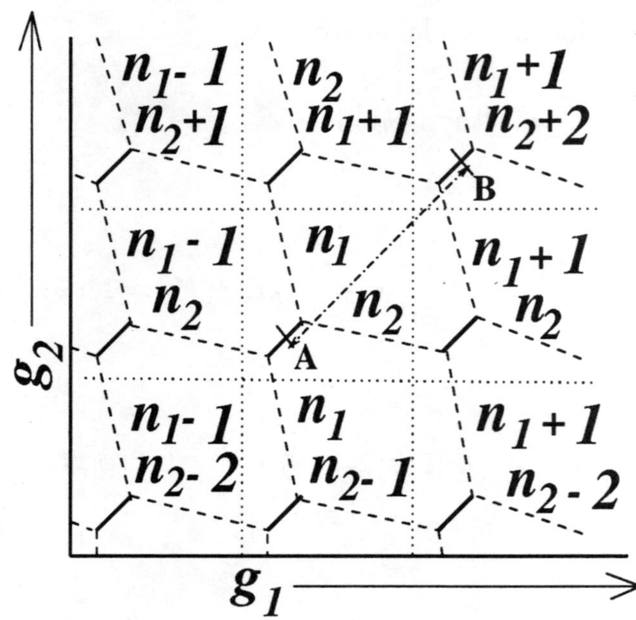

When J is finite but large, we are in the *Weak tunneling limit*. Here, the peaks split. The square grid (in the g_1, g_2 plane) in case of decoupled dots goes over to a hexagonal grid due to the splitting of conductance maxima when $1/J \neq 0$.

The other limit occurs when $J = 0$. In that limit, we have a single big dot. If the two dots were of equal size, one would now have half as many peaks, as we did for the $J = \infty$ case. When $J \neq 0$, but is small, then we are in the *strong tunneling or weak barrier limit*.

As we go from weak tunneling to strong tunneling, the Coulomb blockade peaks start moving nearer, but pairs start receding, until the Coulomb blockade peaks which have moved nearer coalesce into one. That is how, the number of peaks become half in the strong tunneling limit.

Now the idea is to use LL theory and the effective action to quantitatively obtain the relation between peak splitting and barrier conductance. The calculational details are given in [6]. Here, we only quote the results.

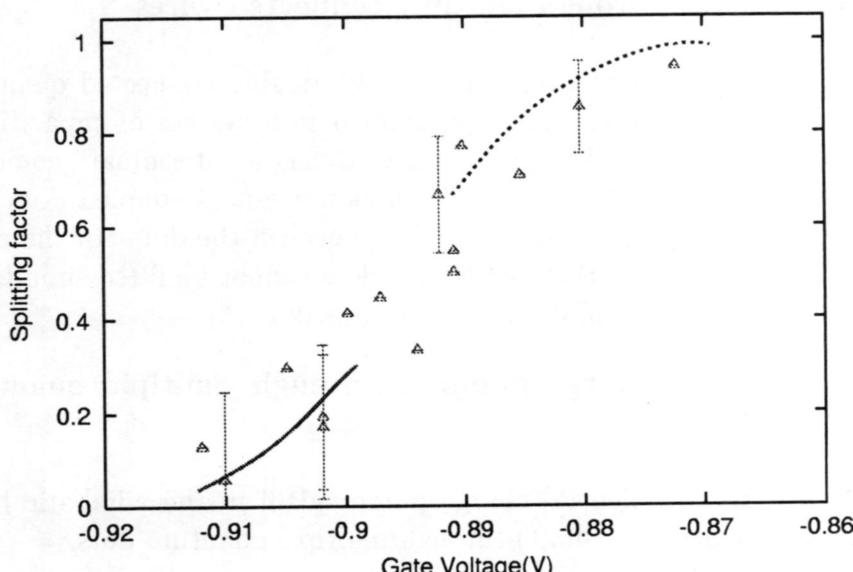

The splitting factor S is defined in terms of $S = \frac{2\Delta V_S}{\Delta V_P}$, where δV_S is the splitting of the peaks and ΔV_P is the distance between pairs of peaks. In the weak tunneling limit, (large J),

$$\Delta V_S = \frac{U^2}{8J} \quad \text{and} \quad \Delta V_P = U$$

and in the strong tunneling limit, (small J),

$$\Delta V_S = \frac{U}{2} - 2J \quad \text{and} \quad \Delta V_P = U \ .$$

We compare this with the experimental data given in [5]. A χ^2 fit gives a goodness of fit of 85% and 83% for weak and strong tunneling respectively. In fact, this is quite good as compared to 75% and 81% for a more complicated theory by Golden and Halperin[10].

We have also predicted a power law dependence on dot sizes. But there are no experiments yet, where the sizes of the dots are varied.

• Transport through multiply connected wires

We also studied transport through doubly connected quantum wires[11]. This could also be mapped to a system of capacitively coupled quantum dots in parallel. It has a 4-terminal geometry. Here again, we can study the conductance of decoupled dots and then allow for a capacitive coupling between the dots. In the presence of coupling, CB through both dots cannot be lifted simultaneously and we get splitting of the CB peaks.

• Quantised charge pumping through multiple quantum dots

We also studied[12] charge pumping[13] in the adiabatic limit (small frequency ω limit) through multiple quantum dots.

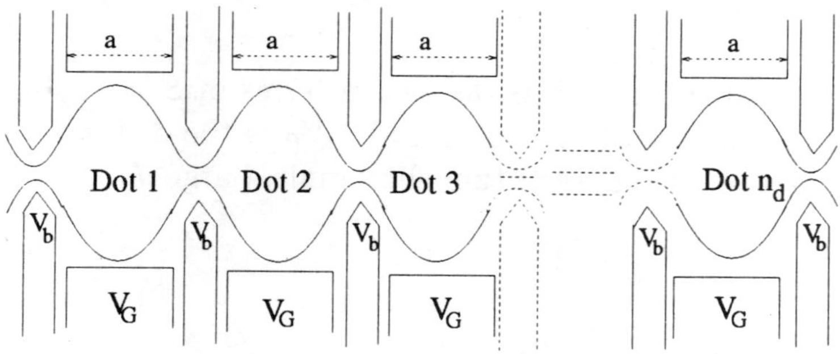

n_d dots are defined on a two dimensional electron gas. The barriers forming the dots are denoted as V_{bi} and the gate voltages controlling the density in the dots are denoted as V_G. The shape of the dots is varied by periodically modulating V_{bi}. Quantisation of the charge that is pumped through the system is expected for a perfectly periodic system[14]. However, we find approximate quantisation even for a small number of dots.

CONCLUSION

The main point that we wish to make here is that transport through quantum dots, quantum wires, and in general through mesoscopic systems, is a fascinating field with many open questions. In fact, it is a general challenge to include interactions in other mesoscopic phenomena as well. Many of the phenomena are disorder driven. Hence, in this field we have the problem of including disorder, interactions and surface effects all at once.

References

[1] For an introduction, see S. Datta, *Electronic transport in mesoscopic systems* (Cambridge University Press, Cambridge, 1995).

[2] *Single charge tunneling*, edited by H. Grabert and M. H. Devoret, (Plenum Press, New York, 1992).

[3] P. Nozieres, Jnl of Low Temperature Physics, **17**, 31 (1974).

[4] D. Goldhaber-Gordon *et al*, Nature **391**, 156 (1998).

[5] F. R. Waugh et. al., Phys. Rev. Lett. 75 (705); F. R. Waugh *et al*, Phys. Rev. **B53**, 1413 (1996).

[6] Sourin Das and Sumathi Rao, Phys. Rev **B68**, 073301 (2004).

[7] K. A. Matveev, Phys. Rev. **B51**, (1995).

[8] P. Rojt, Y. Meir and A. Auerbach, Phys. Rev. Lett.**89**, 256401 (2002).

[9] For a review, see S. Rao and D. Sen, in *Field Theories in Condensed Matter Physics*, edited by S. Rao (Hindustan Book Agency, New Delhi, 2001).

[10] J. M. Golden and B. I. Halperin, Phys. Rev. **B53**, 3893 (1996); *ibid* **65**, 115326 (2002).

[11] Sourin Das and Sumathi Rao, cond-mat/0310713.

[12] Argha Banerjee, Sourin Das and Sumathi Rao, cond-mat/0307234.

[13] D.J. Thouless, Phys. Rev. **B27**, 6083 (1983);Q. Niu, Phys. Rev. **B34**, 5093 (1986); P. Brouwer, Phys. Rev. **B58**, R10135 (1998).

[14] D.J. Thouless, Phys. Rev. **B27**, 6083 (1983).

Physics of Solids, Nuclei and Particles
Editor: R. Sahu

Physics of Two Terminal Semiconductor Devices for High Frequency Generation

S.P. Pati

Department of Physics, Sambalpur University, Jyoti Vihar 768 019
Sambalpur, Orissa, India

ABSTRACT: *The present Day Electronics Communication mostly carried out with high frequency in the range of Microwave to MM-Wave. Miniaturization of Electronics System may only be possible through use of Solid State Devices for the generation of this range of frequency. A comprehensive account of Two Terminal Solid State Devices capable of high frequency generation is presented. The scope includes the Physics of Gunn, Tunnel and Impatt diodes mostly operate with exhibition of high frequency negative resistance generation.*

Introduction: The rapid Development of Electronics and Computer Science during last two decades, covering the fourth generation has evolved the new era of IT. For successful growth of IT in the service of Society, the present Day Electronics Communication system has been modulated to use higher and higher BW of frequency for accommodating more and more number of channels, which in turn enhances the efficiency of the system. The frequency range from 10 GHz to 300 GHz is being explored. The conventional microwave tubes are usually incapable of producing such high frequencies. Further the giant size of microwave generator causes inconveniences for designing of Modern Electronics System. Thus Solid State Generators mostly in the form Two Terminal Devices are used these days for microwave/mm-wave generation. The most effective way to design RF Solid State Devices is through negative resistance generation following different physical principles of Semiconductor.

Various Two Terminal Solid State Devices suitable for generation of microwave/mm-waves have been discussed through this paper. The Gunn device operating through Gunn effect, the Tunnel and other allied devices producing rf negative resistance through tunneling process and Impatt diodes exhibiting generation of negative resistance through combined phenomena of transit time delay and avalanche phase delay are mostly presented.

The general physical principle of operation of the devices with the art of generation of rf would be narrated. Different moods of operation and some state-of-art results would also be presented giving a comprehensive idea of such device performance. The devices appear in this paper in usual sequence.

A. GUNN DIODES:

In group III-IV compound semiconductor like GaAs, InP etc., the E-k diagram show direct band gap i.e., the maxima of valence band lie over the minima of conduction band thus making it possible for direct transition of electrons from Valence band to conduction band. However in these Semiconductors, there exists Satellite Valley in conduction band, which initiates the phenomenon of negative differential resistance/ mobility (NDR) region in the form Gunn effect. A bulk of Semiconductor drawn with Two Terminals when subjected to variable Voltage across the terminals, produces current flow through the device. It is expected to give an Ohmic variation of I-V characteristics. However it is interesting to see that within a specific range of Electric Field (Voltage), the I-V plot shows a NDR region. This region can be exploited for generation of rf Gunn Oscillation. The Physics behind these phenomena can be explained from the following analysis. The Band Diagram in III-V semiconductor (say n-GaAs) is shown in Fig. 1.

The E-k diagram shows multiple valleys in the Conduction Band, the first (lowest) one corresponds to k=0 making the Valley minimum just above the Valance band maximum just making it a Direct Band Gap Semiconductor. The band gap for the first valley is

1.43 eV. The satellite Valley positioned off the k=0 line and the band energy difference between these two valleys remain around 0.31 eV.

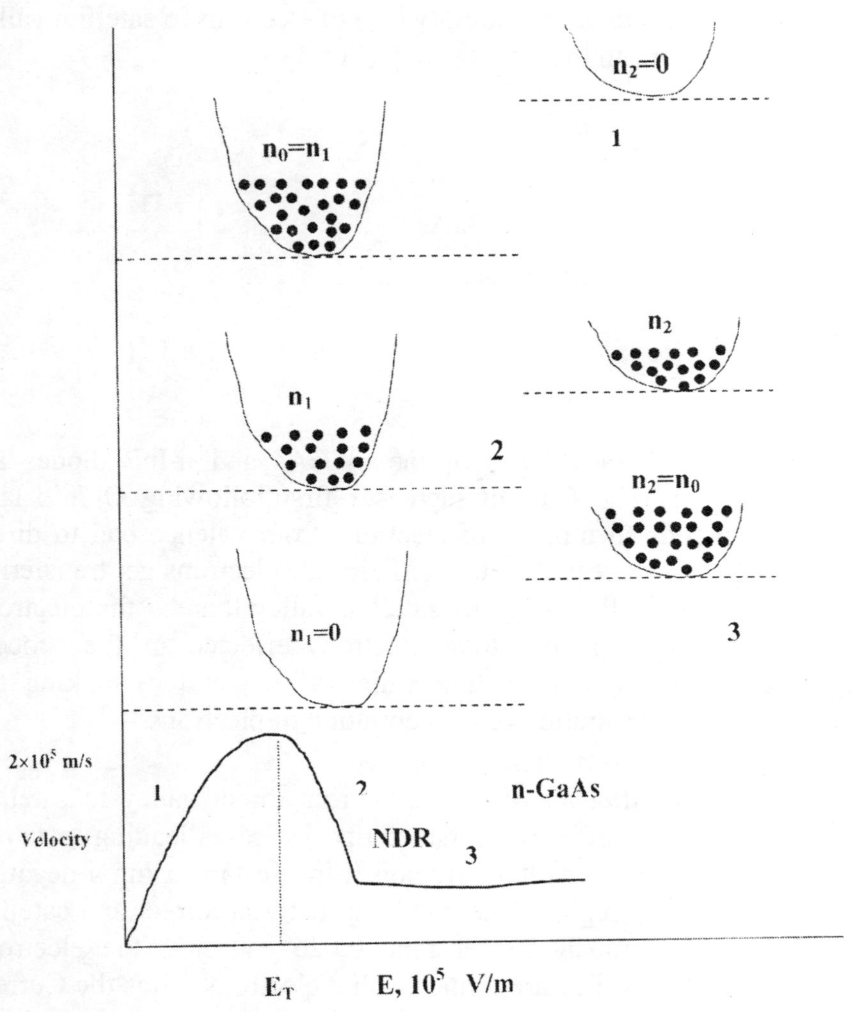

Fig. 1: Electron distribution in different valley and I-V characteristics

On supply of Bias across the n-GaAs, the energized electrons from the Valence Band would start elevating to the Conduction Band, first occupying the states in the First Valley due to direct transition. On subsequent increase of the field across the terminals, the electrons from the first direct valley would transfer to the satellite valley through climbing the energy barrier of 0.31 eV. The electron mobility in the direct valley (μ_1) & the effective mass m_2 in satellite valley are respectively greater than the mobility (μ_2) of electrons in satellite valley & effective mass m_1 in the direct valley (Fig. 1).

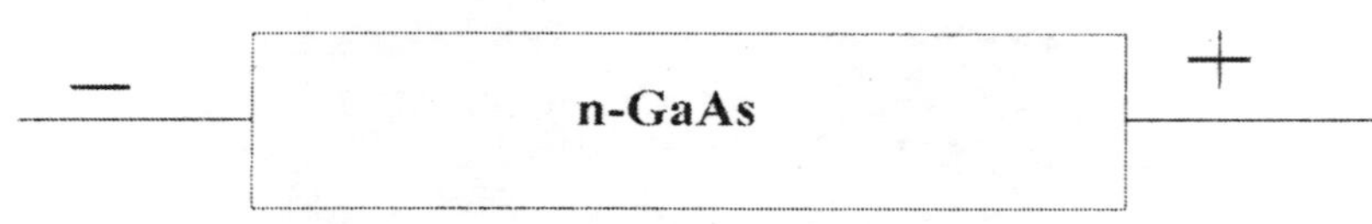

The I-V characteristics of the n-GaAs and n-InP diodes are shown in Fig. 1. The Current increases first following Ohm's law, which corresponds to transfer of electrons from valence bad to direct valley. On further increase of electric field, the electrons get transferred from lower direct valley to upper satellite valley thereby the electrons losing the mobility. If n_0 total electrons effected in the process distributed in Direct and satellite valley as n_1 and n_2 making the conductivity and distribution of concentration of electrons,

$$\sigma = n_1 \mu_1 + + n_2 \mu_2 \text{ and } n_0 = n_1 + n_2$$

As the process of transfer of electrons from direct valley to satellite valley continues, the effective conductivity decreases leading to fall in current with increase of voltage (region II in Fig.1) causing a negative differential mobility region. The band gap between direct and satellite valley 0.31 eV is much greater than 0.026 eV, thus the electrons jumping to satellite valley are termed as hot electrons. Thus the Current first increases in region I due to transfer of electrons from valence band to conduction band, decreases in region II due to loss of mobility of transferred electrons from direct to satellite valley and again increases marginally to attain near saturation in region III after all electrons transferred to satellite valley. Table 1 shows the band gap energy, critical field etc. in GaAs, InP and CdTe materials. This effect of exhibition of NDR is known as Gunn effect and the theory to explain such mechanism is presented in RWH theory, which is beyond the scope of this paper.

Table I

Material	Band Gap Energy In eV	Diff in energy level Between Valley	Field for Peak Velocity (V/m)
GaAs	1.43	0.36	3.2
InP	1.26	0.60	10.5
CdTe	1.44	0.51	13.0

There are several modes of production of rf oscillations in Gunn diode based on the properties of the III-V Semiconductors. A charge domain or fluctuation formed grows in NDR region till region III is reached and then the formation/growth repeated. For InP the difference in Inter Valley Energy in Conduction band is 0.6 eV, which is higher than GaAs, which ensures faster electrons/shorter TT/higher frequency, and possible higher efficiency. The experimental report shows generation of 50 GHz (fundamental mode) and 100 GHz in harmonic mode for GaAs where as for InP the frequency can be pushed to 150 GHz. The power produced in GaAs diode is reported to be 168 mW at 94 GHz. The InP diode has been reported to produce 220 mW at 9.5% efficiency for 10 GHz generations. Some new materials like InGaAs, AlGaAs have been explored but can be used for low f operation. As per extrapolated prediction Gunn oscillations can be generated unto 300 GHz. The research work now continues in the areas of enhancement of frequency, search for new materials, compositional variation of composite semiconductor and new analytical models to represent the Gunn diode operation in a full-fledged manner.

B. TUNNEL DIODES:

In 1958, Esaki proposed a Nobel work by showing exhibition of NDR in heavily doped p-n junction due to tunneling of charge carriers. In this case the impurity doping on either side of the junction is very heavy to the order of 10^{19} to $10^{20}/m^3$, which makes the diode depletion zone to be within 100 A^0. Thus the Fermi level goes into the conduction band/valence bands. Under no bias condition, the filled states in valence band and empty states in conduction band remain mismatched. As the junction is subjected to progressive forward bias, the current first increases, then as the filled band and the empty band approach each other and starts facing each other, the tunneling of electrons would take place from n-side to p-side (Fig. 2). The tunneling probability increases and becomes maximum when the filed and the empty bands face with 1:1 correspondence. The Current increase. On further rise of forwarding voltage, the empty band and the filled band again gradually mismatched causing less number of electrons to tunnel from n to p. Thus the current decreases on rise of forward bias, which in turn exhibits NDR region. The current reaches a minimum point. When the forward bias is further increased the current increases

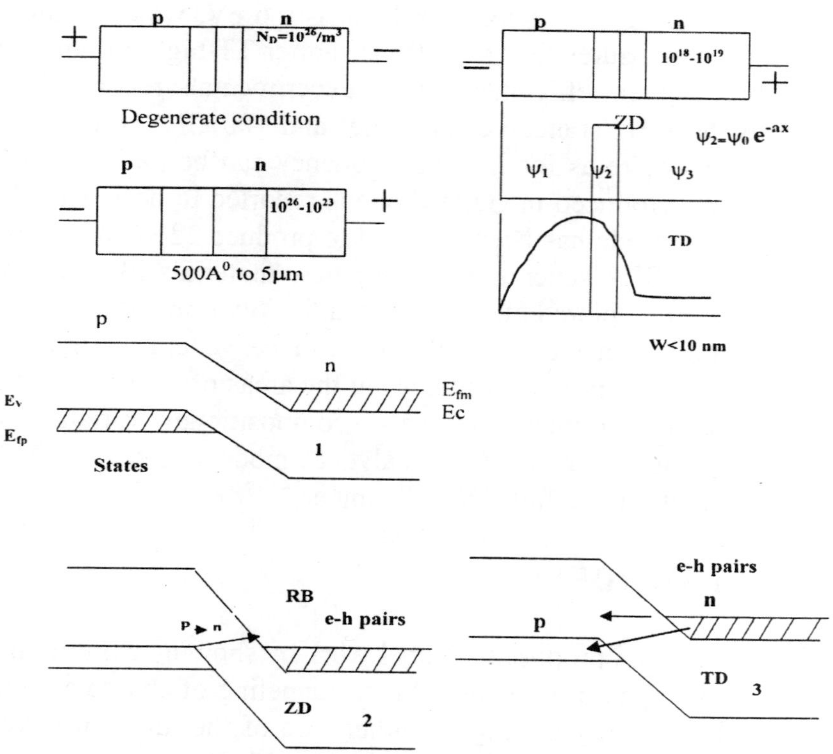

Fig. 2:   Band diagram in Tunnel Diode

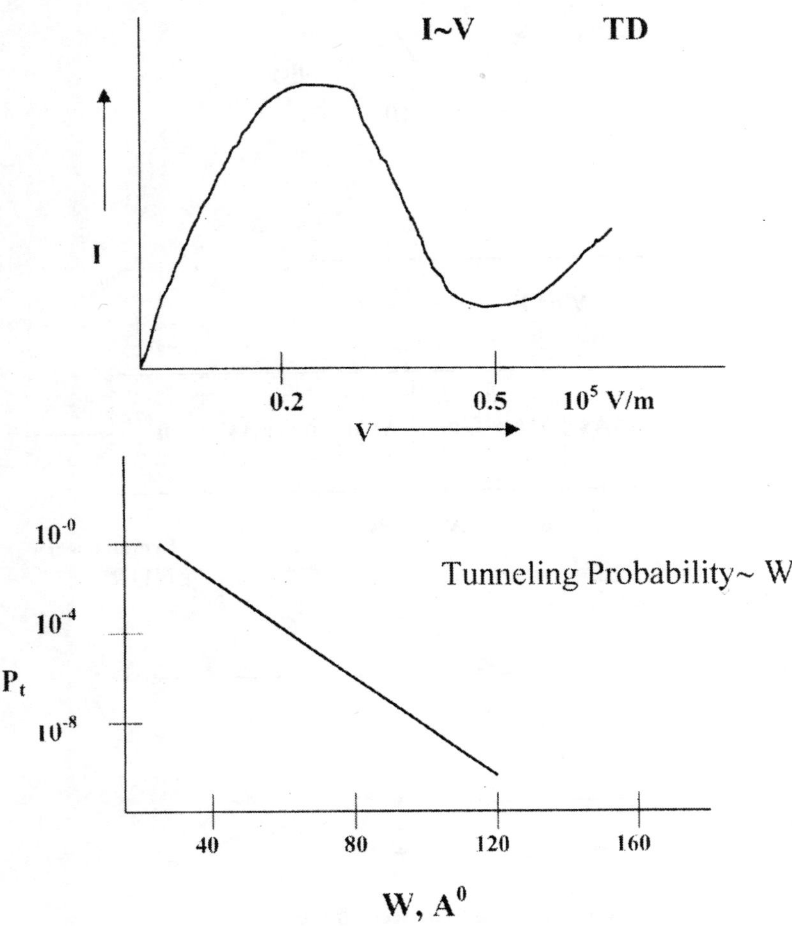

Fig. 3: I-V chararacteristics and $P_t \sim$ W in Tunnel diodes

following diode equation due to injection of electrons from n to p and holes from p to n. When the bias is made in the reverse direction, the tunneling of charge carriers is possible by the minority carriers. These carriers multiply to cause Zener Breakdown.. Fig.2 shows the band diagrams under various biasing conditions and corresponding I-V characteristics in Fig. 3. The region II gives the NDR region, which can be utilized for generation of microwave oscillations. The tunneling

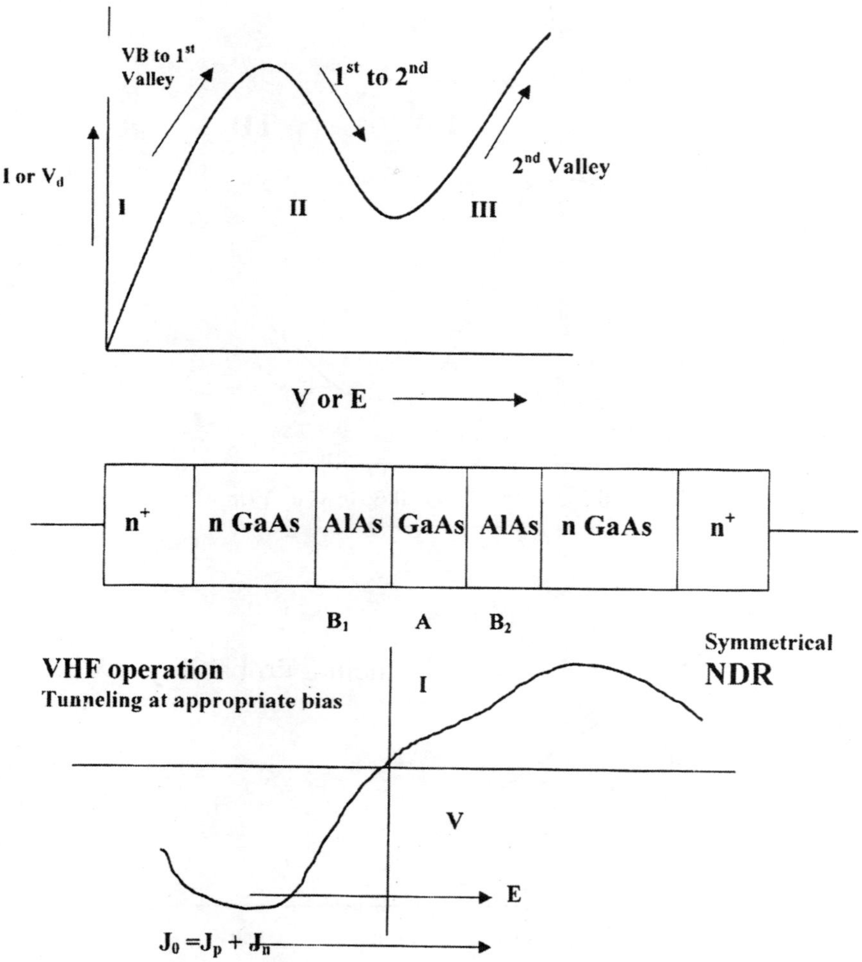

Fig. 4: Resonant Tunneling devices RTD- Double Barrier

probability increases with narrowing of depletion zone as shown in Fig. 3. The tunneling probability depends on the type of barrier and can be computed from usual formula. The variation of Tunneling Probabilty with with width of depletion zone is also shown in Fig. 3. The tunneling is nearly zero for depletion zone extending beyond 160 A^0 The equivalent circuit of a tunnel diode can be shown to have a parallel combination of Cs (Junction storage capacitance) and -Rn (negative resistance). The negative resistance for a standard diode may be around 30 ohms and the value Cs is around 20 pf.. The package for the tunnel diode in cap form would be equivalent to Series Positive resistance (Rs) and the package inductance Lp. The optimum frequency of oscillation from the tunnel diode is given as fm= Sqrt((Rn/Rs)-1)/2π RnCs. The advantages of tunnel diode are mostly due to low generation noise. However the power and efficiency would also be low as the tunnel diode operates with little forward bias.

The recent trend is the design and use of Resonant Tunneling Devices. The I-V characteristics of this device are shown in Fig. 4. Double Barrier mode of tunnel diode can push the frequency of operation. The structure is n+(GaAs)-AlAs-GaAs-AlAs-(GaAs)p+ where tunneling can take place from either barrier. Alternate GaAs and AlGaAs layers generate a super lattice structure where NDR region can be exhibited through tunneling. The Power realizable from Tunnel diodes in any mode of operation remains low as the NDR region occurs within low voltage region. However the noise generation always remains within tolerable limit, as the tunneling is a regular process. The efficiency becomes moderate upto 100 GHz.

C. IMPATT DIODES:

The other two terminal solid-state devices, which is regarded as the premier device for microwave/mm-wave generation, is Impatt Diode which stands for abbreviation Impact Ionization Avalanche Transit Time. There are several advantages for this device over its counterpart as described earlier. The entire range of Micro/MM Wave regions even generation of lower Sub-Mm Wave is possible by this device. The device can be realized from any Semiconductors like Elemental (Si/Ge), Binary (GaAs, GaP, InP etc.), Ternary (GaInP etc.) and Quaternary (GaInAsP) as there are no specific semiconductor properties on which device operation critically depends. The device is a reverse junction of any structure and thus the device operation can be modulated on device structure modification. The wide-ranging operating condition like the diode current and temp is acceptable for diode operation, which may enable to realize high power. The device efficiency is comparatively very high, may become upto 25% even for mm-wave operation. The only disadvantage is the production of high rf avalanche noise.

The first proposition of Impatt mode rf oscillation came in 1958 from W. T. Read from Read diode structure of the form n+pip+. Then in 1965, Lee et al and Johnston et al reported rf oscillations in X-band from simple p-n junction. After 1965, several groups in India and Abroad carried out research activities (both theoretical and experimental) on the Physics, Property, Design and realization of this Device through advanced technology. The semiconductor parameters as relevant to Impatt operation also measured experimentally to support theoretical research. Works are also carried out towards enhancement of Frequency, Power Output and Efficiency and also to explore new semiconductors with homo/hetero junctions. A reverse biased p-n junction under avalanche breakdown can exhibit existence of rf negative resistance, which can be exploited to generate high frequency oscillations.

The basic phenomena involved in rf generation are Impact Ionization, Saturated Drift Velocity, Avalanche Breakdown, Transit Time and Avalanche Phase Delay. A brief explanation of each physical phenomenon is outlined below.

IMPACT IONIZATION:

Diode under reverse bias is subjected to a reverse electric field, which acts as forward field for minority charge carriers causing them to cross over the junction and generate a reverse current. These carriers on entering the depletion zone gains energy till it meets a collision with another lattice point. The collision may emit lf phonons, optical phonons at the first instance respectively beyond threshold energies of ε_l and ε_{op}. When the gain in energy between collisions is beyond the ionization threshold, ε_i , the target atoms get ionized on production of additional e-h pairs. Such generation of e-h pairs are measured by electron or hole ionization rate as per its initiation. The electron ionization rate α_n, and hole ionization rate α_p are measured as number of ionizing collisions per unit distance (/m). The ionization rate depends on electric field, temperature, material, crystal orientation etc.

The ionization rate are measured experimentally through photo multiplication and other sophisticated methods. The experimentally determined values of ionization rate vs. electric field across depletion zone can be made correspondence to an exponential function of following form,

$$\alpha_{n,p} = A_{n,p}\exp(-b_{n,p}/E)$$

Where $A_{n,p}$ and $b_{n,p}$ are constants whose values can be obtained through computer aided curve fit in technique and may be different for different ranges of field, different semiconductors and other material/operating conditions. The ionization rate increases following the exponential function and tends to attain saturation at very high values of electric fields. It can become as high as beyond 10^6 near the field maximum. These rates have been measured for several semiconductors of all varieties and reported in literature. The ratio of $\alpha n/\alpha p$ is not normally unity and may become more than 1 (for Si etc.) and less than 1 (for InP, GaAs etc.). This has made the theoretical research of more in depth nature.

The charge carriers get multiplied within the depletion zones i.e. electron from saturated value (Jns) at p edge to maximum Jn at n edge. The ratio Jn/Jns=Mn and diode breaks down under avalanche multiplication process when Mn $\cong\infty$ ideally, and beyond 10^6 practically.

Saturated Drift Velocity: As the charge carrier enters the depletion zone, its value increases. Finally on attaining a particular value, the velocities become saturated at fixed values which may be called as saturated drift velocity and its value is different for electrons, holes; for different semiconductors; for different operating temperatures etc. Exponential function of form,

Vn,p= $Vs_{n,p}(1-\exp(-\mu_{n,p}/E))$, can be used for computing carrier velocity at particular electric field. However for direct band semiconductors like GaAs, InP etc., the v – E plot shows NDR region for electrons followed by saturated region as usual. When a minimum punch through factor (Minimum electric field at the edges of depletion zone) is maintained, the electric field becomes sufficient for the carriers attaining saturated values along the entire depletion zone.

Transit Time: The carriers on moving with saturated velocity within the depletion zone consumes time which is termed as transit time as the time taken by the carriers for traveling the entire depletion zone ($\tau = L/V_s$). The width of the diode (L) thus fixes the frequency of operation. Further when a charge bump formed at the left edge due to imposition of ac field, crosses the active zone to take time τ, it includes transit time phase delay depending on frequency and for optimal Impatt action, the situation is modulated to make it $\Pi/2$.

Avalanche Phase Delay: The generation rates of charge carriers depend on ionization rate, carrier concentration and the carrier velocity. The expression for this can be given in the form of,

$$G = \alpha_n \, V_n n + \alpha_p \, V_p \, p$$

When an rf voltage is assumed to be superimposed on the dc breakdown field, the ionization rate follows the rf cycle thus peaks at peak of rf voltage. However the growth of carrier concentration lags behind the rf voltage by suitable angle because n and p depends on $\alpha_{n,p}$, $V_{n.p}$, rf cycle. This delay is named as avalanche phase delay, which is the phase lag, exhibited between rf cycle and growth of charge bump. This phase delay together with transit time delay if becomes $\Pi/2$ to $3\Pi/2$, the ratio of rf voltage to current becomes negative leading to generation of rf negative resistance at the frequency of imposed rf voltage.

As have been indicated earlier, any diode structure can exhibit negative resistance. A schematic diode structure is shown in Fig. 5. The figure is self-explanatory and identifies the central avalanche zone separating the two drift zones (DDR Structure), one for electron and the other for holes. Electrons as minority carriers enter from p to n side, gets multiplied in the avalanche zone and the full electron current reaches the n side edge. If either doping of n or p side becomes of the order $10^{26}/m^3$ or so the structure becomes a SDR where only one drift region exists. The doping of n and p regions becomes flat (Constant) for flat SDR/DDR. The doping can be modulated to make it multi layered by introduction of a charge bump of high impurity concentration near the junction plane, like Low-High-Low and so on. The flat SDR/DDR provides small efficiency say 15% for X-band, 10% for W-band. The efficiency is enhanced in modulated low-high-low structure to 25% for X-band. The author has even attained closest to theoretically optimum efficiency of 28% by introduction of two bumps on

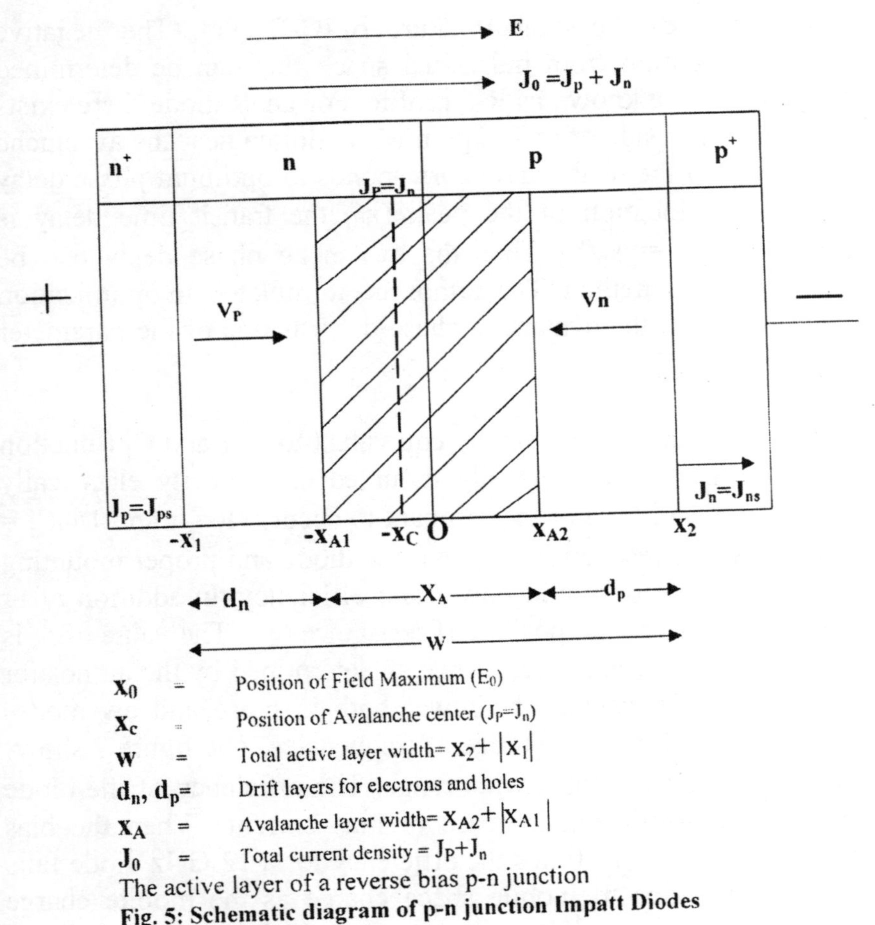

X_0	=	Position of Field Maximum (E_0)
X_c	=	Position of Avalanche center ($J_p = J_n$)
W	=	Total active layer width= $X_2 + |X_1|$
d_n, d_p	=	Drift layers for electrons and holes
X_A	=	Avalanche layer width= $X_{A2} + |X_{A1}|$
J_0	=	Total current density = $J_p + J_n$

The active layer of a reverse bias p-n junction

Fig. 5: Schematic diagram of p-n junction Impatt Diodes

either side of junction. The Read diode is also promising structure as the efficiency is high.

The diodes can be analyzed/optimized following techniques as proposed by author through sophisticated computer program for dc/rf analyses. The diode equations are solved by double iterative programs subjected to usual boundary conditions under different conditions of analysis. The dc analysis fixes the edges of diode layer accurately, gives the extension of avalanche/drift zones, voltage drops and efficiency. The rf analysis can give the BW of rf oscillation, G-B plot, quality factor etc.

The figure 6 gives the typical nature of G-B plot. The negative resistance contribution from individual space step can be determined and the typical curve known as R-x profile. For DDR diode there exists two peaks on either side of drift region with minima near the avalanche zone. The peak in the drift region corresponds to optimum phase delay of Π. From the location of the peak (x_p) the transit time delay is calculated from $\tau_t = x_p/V$. Thus the avalanche phase delay can be computed from this method. The author has formulated an optimization technique for which the avalanche phase delay is one of the parameter (Optimum value $\Pi/2$.

The diode thus is electrically equivalent to a $-r$ and Cj (junction capacitance). The diode is usually mounted on a cavity electrically equivalent to +R and Lc. For the resonant to occur, $+R= -r$ and $1/\omega Cj = \omega Lc$. The design of resonant cavity and the diode and proper mounting can make it possible optimum power and efficiency. In addition of $-r$ the diode also can generate positive rf resistance (r_s). The value of r_s is almost one tenth of negative resistance as determined by the author for several cases. Experimentally rf Power both in pulse and cw modes have been reported for a wide range of frequencies. The figure 7 shows the plots of P-f and η-f for a wide range. The efficiency of the diode falls with diode width and increase of bias current. When the bias current is increased by 10 times the efficiency for a 12 GHz diode falls from 15% to 1% due to mobile space charge as the mobile charge concentration becomes comparable to doping at high bias current.

The present study includes the possible generation of rf in a non-conventional diode having the structural form $p + n \gamma np +$ or $n + p \pi$

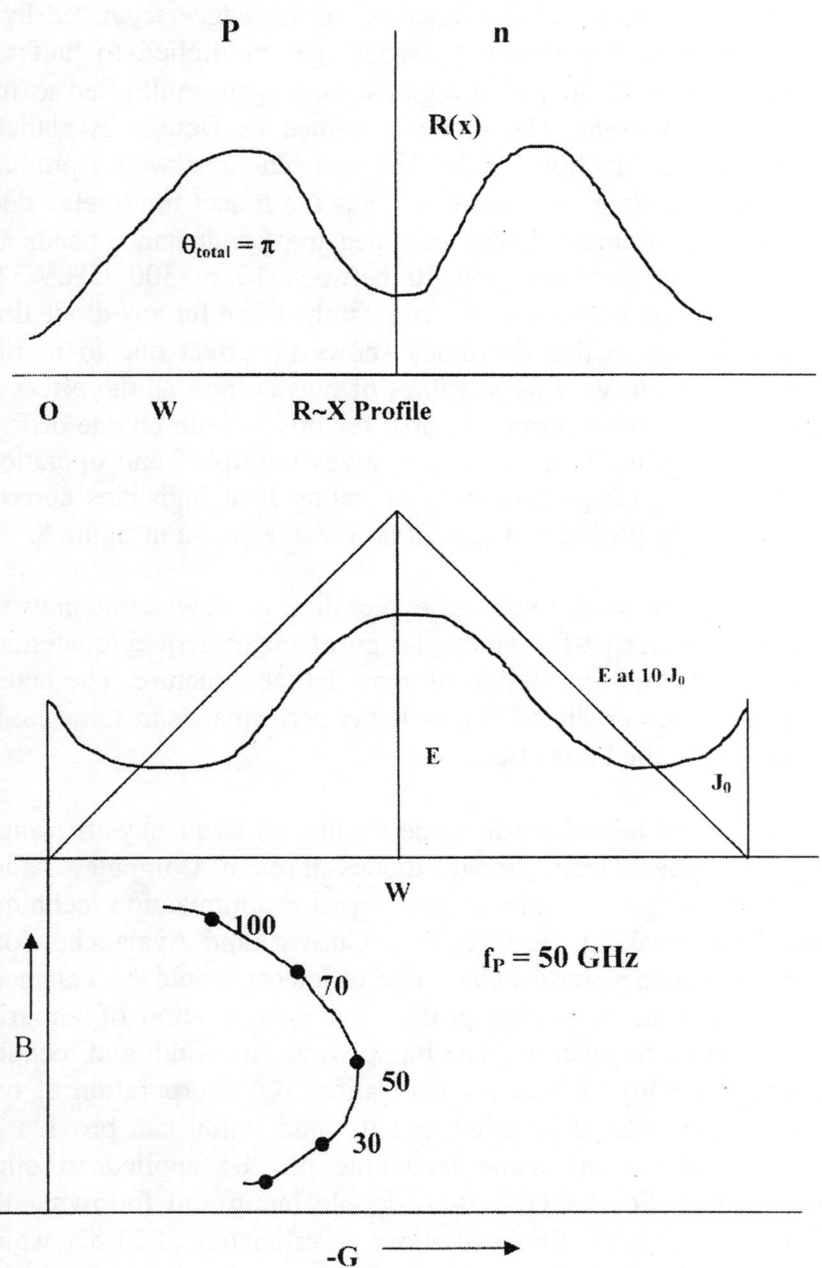

Fig. 6: R~X Profile, Field profile and G-B plot of tropical Impatt Diodes

np+. In this case there are two junctions at the edges separated by a single drift region. The minority carriers get multiplied to half the maximum value, drifts in γ or π regions, then again multiplied to full value of carrier current. The diode is named as Double Avalanche Region or Double Junction Diode. The two junctions would produce avalanche delay adding to close of π. Thus the transit time delay does not become a requirement. In this case negative conductance bands are obtainable in few numbers may 20 between 10 to 300 GHz. The numbers of frequency bands remain almost the same for any diode drift region. In addition to this the diode shows no effect due to mobile space charge even at very large values of bias current as the effect of mobile charge get compensated in drift regions as both charge drift in common drift region. This diode thus gives multiple band operation, high efficiency, and high power by operating it at high bias current. The structure, field profile and current profile are shown in figure 8.

The present study includes exploration of new semiconductor materials with better performance, design of binary/ternary/quaternary homo/hetero junction and design of super lattice structure. The hetero junctions have been predicted to give better performance in some cases. The author has studied this effect.

The author has taken up some studies on frequency/efficiency/ power enhancement from Impatt diodes through Computer Aided Analysis and Design. A sophisticated stepwise optimization technique has been framed taking the PTF, Bias Current, and Avalanche Zone width and avalanche phase delay. The efficiency could be enhanced through modulation of doping profile and incorporation of impurity bumps at or close to junction. The bump width, its width and location can be varied for high efficiency realization. By incorporation of two bumps on either side of junction and its modulation can provide an efficiency of 28%. The same technique can be applied to other frequency bands. For 94 GHz with double bump and following the optimization process as stipulated above an efficiency of 21.8% which if realized would be a commendable performance.

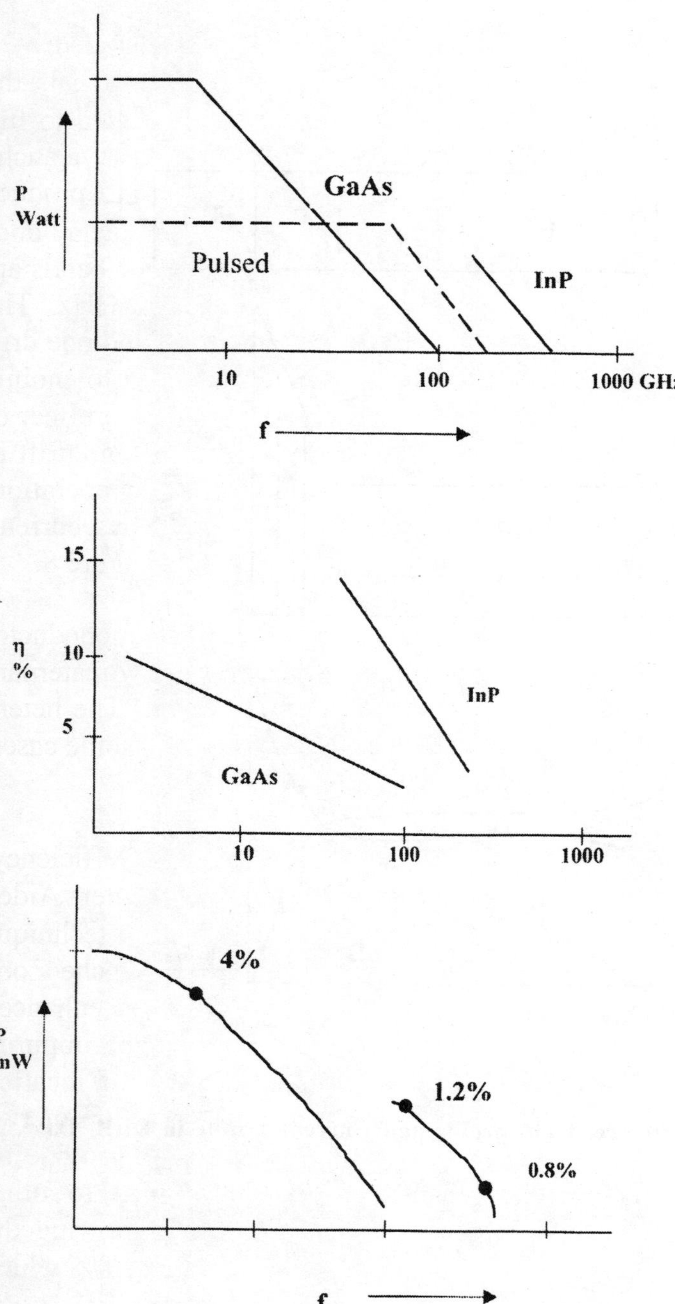

Fig. 7: P- f and η -f of reported Impatt Diodes

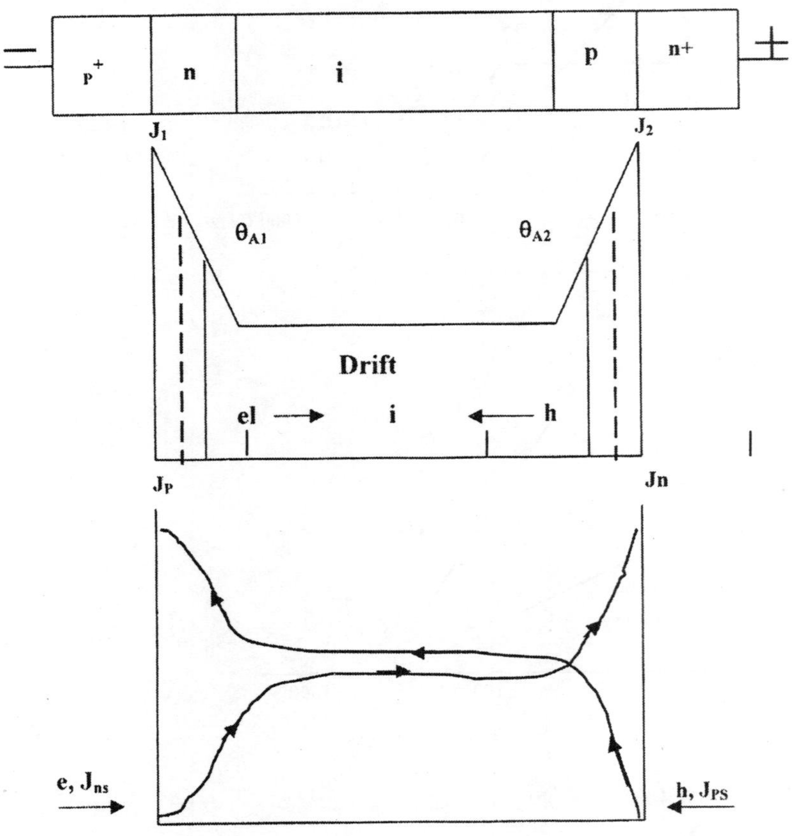

Fig. No. 8: Structure, Field profile and Current profile in DAR/ DJ Impatt Diodes

Pushing the power input through enhancement of bias current can increase the power output. The increase of bias current also pushes the frequency by sqrt(J) law. The performance deterioration can be compensated by suitable technique. Further the frequency of operation can be pushed by operating the diodes in harmonic mode. Thus amongst all solid-state devices capable of generating microwaves, the Impatt diode has a large number of advantages. The microwaves/mm-waves find use in present day electronics communication systems and thus the above referred devices find a lot of applications for point to point to communication accommodating a large number of channels. Since the scope of this paper is limited both as far as contents and space, the author wishes to put an end to the paper here.

Physics of Solids, Nuclei and Particles
Editor: R. Sahu

A Microscopic Model for Valence Transition in Yb- and Eu- Based Systems

G.G. Reddy[†], A. Ramakanth[†] and S.K. Ghatak[‡]

[†]Department of Physics, Kakatiya University, Warangal 506 009, India
[‡]Department of Physics, Indian Institute of Technology, Kharagpur 721 302, India

Abstract

Motivated by the experimental studies on the temperature- and field- induced valence transitions in Yb- and Eu- based rare earth compounds, a model calculation is performed. For this purpose, the so called extended periodic Anderson model(EPAM) is used. The model is solved first by making a mean field approximation to the Falicov-Kimball term and then taking the intra site Coulomb interaction strength to infinity. The resulting coupled equations for the spin dependent average occupations of the localized and itenerant states and the excitonic correlations have been solved self-consistently for different constellations of model parameters. The model calculations show that it is possible to induce valence transitions in the system by either varying the temperature or the field or both. The results obtained bear a rather striking similarity to the experimental observations not only in general but also in most of the details.

1 Introduction

The phenomenon of intermediate valence continues to attract the attention of both experimentalists and theorists. This phenomenon is observed in rare earth metals alloys and compounds. In these systems, the ground state energy of two valence states of the rare earth ion, eg., Yb^{2+} and Yb^{3+} or Eu^{3+} and Eu^{2+} are close to each other. So, under suitable conditions, the rare earth ion fluctuates between the two valence states and again when an external parameter such as temperature, pressure or magnetic field is varied, it makes a transition from the fluctuating valence or intermediate valence (IV) state to a stable valence state. The two valence states in question have unique magnetic properties for all the rare earth ions. One of them is nonmagnetic ($J = 0$) while the other is magnetic ($J \neq 0$). Taking again the above cited examples, Yb^{2+} has $J = 0$ and Yb^{3+} has $J = 7/2$ whereas Eu^{3+} has $J = 0$ and Eu^{2+} has $J = 7/2$. As a result, the transition from IV state to a stable state is followed by either a metal-insulator transition or a magnetic transition. Therefore, the transition can be tracked either by looking at the valence of the rare earth ion or by studying the magnetic properties of the system. Experimentally, it was found that at least in Yb- and Eu- based systems, the transition takes place from an IV state to a magnetic state[1-21]. The transition is some times continuous and some times discontinuous. Further, it is also possible that for the same system, the transition is continuous when driven by temperature and discontinuous when driven by field. For both Yb- and Eu-based systems, When a phase diagram was constructed in terms of reduced variables, a universality was observed[10,17]. All the experimental observations regarding both the temperature- and field-induced valence transitions were explained using a phenomenological model called the interconfiguration fluctuation model. Even though there is a vast amount of experimental data available, there are not many calculations based on microscopic models which try to simulate the temperature- and field-induced valence transitions. In view of the richness of the experimental data and the lack of a microscopic model calculation specifically aimed at simulating the experimental observations, it is proposed in this paper to report the calculations performed using EPAM.

2　The model and the approximation

In rare earth systems there are two electron subsystems, namely, the delocalized conduction electrons and the highly localized and strongly correlated 4f-electrons. The two subsystems interact via a hybridization interaction so that electrons fluctuate between the the two subsystems under suitable conditions. Such a situation is adequately described by the so called periodic Anderson model(PAM). We add to this model an additional term, the so called Falicov-Kimball term which describes a repulsive intrasite interaction between the localized and the conduction electrons. Including the external magnetic field the Hamiltonian of the model reads as

$$
\begin{aligned}
\mathcal{H} \;=\;& \sum_{i,j,\sigma} \left(T_{ij} - z_\sigma H \right) d_{i\sigma}^\dagger d_{j\sigma} + \sum_{i,\sigma} \left(e_f - z_\sigma H \right) f_{i\sigma}^\dagger f_{i\sigma} \\
&+\; V \sum_{i,\sigma} \left(f_{i\sigma}^\dagger d_{i\sigma} + h.c. \right) + \frac{U}{2} \sum_{i,\sigma} n_{fi\sigma} n_{fi-\sigma} \\
&+\; G \sum_{i,\sigma,\sigma'} n_{fi\sigma} n_{di\sigma'}
\end{aligned}
\tag{1}
$$

$d_{i\sigma}(f_{i\sigma})$ and $d_{i\sigma}^\dagger(f_{i\sigma}^\dagger)$ are the annihilation and creation operators for the conduction(localized) electrons respectively. $n_{di\sigma} = d_{i\sigma}^\dagger d_{i\sigma} (n_{fi\sigma} = f_{i\sigma}^\dagger f_{i\sigma})$ is the number operator for the conduction (localized) electron. $T_{ij} = \frac{1}{N} \sum_{\mathbf{k}} e^{-i\mathbf{k}\cdot(\mathbf{R}_i - \mathbf{R}_j)} \epsilon(\mathbf{k})$ describes the hopping of conduction electrons (assumed to be in s-state) from site $\mathbf{R}_i$ to site $\mathbf{R}_j$. $\epsilon(\mathbf{k})$ is the Bloch energy. e_f is the position of the localized level measured relative to the centre of gravity of the Bloch band, which is taken to be the zero of the energy scale. U is the Hubbard type intra-atomic Coulomb repulsion between localized (f-) electrons. The hybridization interaction, which represents the fluctuation of an electron between f- and d- states has the strength V and is assumed to be independent of the wave vector $\mathbf{k}$. The external field H couples equally to the spin of both f- and d-electrons. $z_\sigma = \delta_{\sigma\uparrow} - \delta_{\sigma\downarrow}$ is a sign factor. G is the strength of the repulsive interaction between the electrons in the f- and d- states of either spin.

It is clear that the model presented above cannot be solved exactly. We look for the simplest approximation based on physical considerations which is reasonable for IV system. It is well known that intrasite Coulomb interaction between localized electrons is stronger than that between localized and itinerant electrons ($U > G$). The interaction strength G is also small compared to the conduction band width W. So the Falicov-Kimball term is treated in a mean field manner by taking the most general linearization as suggested by Khomskii and Kocharjan[22−26]

$$d^\dagger d f^\dagger f \longrightarrow < d^\dagger d > f^\dagger f + < f^\dagger f > d^\dagger d + < f d^\dagger > (d^\dagger f + f^\dagger d)$$

$$- < f^\dagger f >< d^\dagger d > - < f d^\dagger >< d f^\dagger > \tag{2}$$

When this is incorporated, the Hamiltonian reduces to a renormalized PAM.

$$\tilde{\mathcal{H}} = \sum_{i,j,\sigma} \left(\bar{T}_{ij} - z_\sigma H \right) d^\dagger_{i\sigma} d_{j\sigma} + \sum_{i,\sigma} \left(\bar{e}_f - z_\sigma H \right) f^\dagger_{i\sigma} f_{i\sigma}$$

$$+ \sum_{i,\sigma} \bar{V}_\sigma \left(f^\dagger_{i\sigma} d_{i\sigma} + h.c. \right) + \frac{U}{2} \sum_{i,\sigma} n_{fi\sigma} n_{fi-\sigma} \tag{3}$$

With

$$\bar{T}_{ij} = T_{ij} + G n_f, \ \bar{e}_f = e_f + G n_d \ and \ \bar{V}_\sigma = V + G \gamma_\sigma \ where \ \gamma_\sigma = \left\langle f_{i\sigma} d^\dagger_{i\sigma} \right\rangle \tag{4}$$

$n_f = \sum_\sigma < n_{f\sigma} >$ and $n_d = \sum_\sigma < n_{d\sigma} >$ are the average occupations of the f- and d-states respectively and γ_σ is known in literature as excitonic correlation.

Even after making the approximation given by eqn(2), the model is not exactly solvable. So, we now make a further approximation for the intrasite Coulomb term U. Among all the model parameters, U is the largest and is of the order of $6 − 8eV$ for rare earths. Since U is large it is quite reasonable to take the limit of $U \to \infty$. This type of approximation for U has been used earlier[27−30]. In the limit of $U \to \infty$, the Greens function for the d- and f-electrons and also the mixed Greens function correct up to second order in hybridization (for details see reference [29]) are given by :

$$G_\sigma(\mathbf{k}, E) = << d_{\mathbf{k}\sigma}; d^\dagger_{\mathbf{k}\sigma} >> = \frac{E - \bar{e}_f}{(E - \bar{\epsilon}(\mathbf{k}))(E - \bar{e}_f) - \bar{V}_\sigma^2 (1 - n_{f-\sigma})} \tag{5}$$

$$F_\sigma(\mathbf{k}, E) = \;<< f_{\mathbf{k}\sigma}; f^\dagger_{\mathbf{k}\sigma} >> = \frac{E - \bar\epsilon(\mathbf{k})}{(E - \bar\epsilon(\mathbf{k}))(E - \bar e_f) - \bar V_\sigma^2(1 - n_{f-\sigma})} \tag{6}$$

$$P_\sigma(\mathbf{k}, E) = \;<< d_{\mathbf{k}\sigma}; f^\dagger_{\mathbf{k}\sigma} >> = \frac{\bar V_\sigma(1 - n_{f-\sigma})}{(E - \bar\epsilon(\mathbf{k}))(E - \bar e_f) - \bar V_\sigma^2(1 - n_{f-\sigma})} \tag{7}$$

We have now absorbed the Zeeman term in to the definition of $\bar\epsilon(\mathbf{k}) = \epsilon(\mathbf{k}) + Gn_f - Z_\sigma H$ and $\bar e_f = e_f + Gn_d - Z_\sigma H$.

The equations (5-7) represent our approximate solution of the model. Using the spectral theorem, the spin dependent average occupations $n_{d\sigma}$, $n_{f\sigma}$ and the excitonic correlation γ_σ are given in terms of the respective quasi-particle density of states (DOS) $\rho_{d\sigma}(E)$, $\rho_{f\sigma}(E)$ and $\rho_{df\sigma}(E)$ by

$$n_{d\sigma} = \int_{-\infty}^{\infty} dE \; f(E) \; \rho_{d\sigma}(E) \tag{8}$$

$$n_{f\sigma} = \int_{-\infty}^{\infty} dE \; f(E) \; \rho_{f\sigma}(E) \tag{9}$$

$$\gamma_\sigma = \int_{-\infty}^{\infty} dE \; f(E) \; \rho_{df\sigma}(E) \tag{10}$$

Where $f(E) = 1/\left(1 + e^{\beta(E-\mu)}\right)$ is the Fermi function with $\beta = 1/kT$ and μ is the chemical potential. The quasi-particle DOS are related to the corresponding Green function by[31]

$$\rho_{d\sigma} = -\frac{1}{\pi}\frac{1}{N}\sum_{\mathbf{k}} Im G_\sigma(\mathbf{k}, E + i0^+) \tag{11}$$

$$\rho_{f\sigma} = -\frac{1}{\pi}\frac{1}{N}\sum_{\mathbf{k}} Im F_\sigma(\mathbf{k}, E + i0^+) \tag{12}$$

$$\rho_{df\sigma} = -\frac{1}{\pi}\frac{1}{N}\sum_{\mathbf{k}} Im P_\sigma(\mathbf{k}, E + i0^+) \tag{13}$$

In this particular case, the $\mathbf{k}$-summation can be conveniently replaced by an integration over E using the free Bloch density of states (BDOS):

$$\rho_0(E) = \frac{1}{N}\sum_{\mathbf{k}} \delta(E - \epsilon(\mathbf{k})) \tag{14}$$

$n_{d\sigma}$, $n_{f\sigma}$ and γ_σ have to be evaluated self-consistently. The chemical potential μ has to be fixed by the constraint

$$n = \sum_\sigma (n_{f\sigma} + n_{d\sigma}) = constant \qquad (15)$$

The BDOS is treated as a model parameter.

3   Results and Discussion

As already mentioned, the aim of the present work is to investigate whether field and temperature- induced valence transition can be recovered from the simple valence fluctuation model with a simple and widely used approximation. For that purpose, the equations (8-10) together with eqn(15) have been solved self-consistently using semi-elliptic Bloch density of states in the tight binding approximation[32], for various values of model parameters, as a functions of temperature and magnetic field. The Bloch density of states extends from -0.5eV to $+0.5$eV with the centre chosen as the zero of the energy scale. Throughout the calculations, the band width W is retained as 1eV and the chemical potential is fixed such that $n = 1$. The magnetic field is expressed in eV.

Figure 1 shows the variation of the average f-level occupation n_f as a function of the position e_f of the f-level for various values of G and in absence of magnetic field. As is to be expected, n_f falls as e_f penetrates deeper and deeper into the band. Further, it is reasonable that the variation of n_f becomes sharper with increasing G, becoming almost discontinuous for $G = 0.4$eV. Though not shown in the figure, the size of the discontinuity increases with increasing G. Also, the location of the discontinuity shifts to the right as G is increased.

It was noticed that the critical value of G, at which the transition changes the nature from continuous to discontinuous, increases with the increase in the hybridization strength V. All these results are consistent with the theoretical results obtained by other authors[25-27] using various approximations to solve EPAM. It may be mentioned here that the shifting of e_f in the model is believed

                           Physics of solids, nuclei and particles

to correspond to changing the chemical composition in the experimental sample.

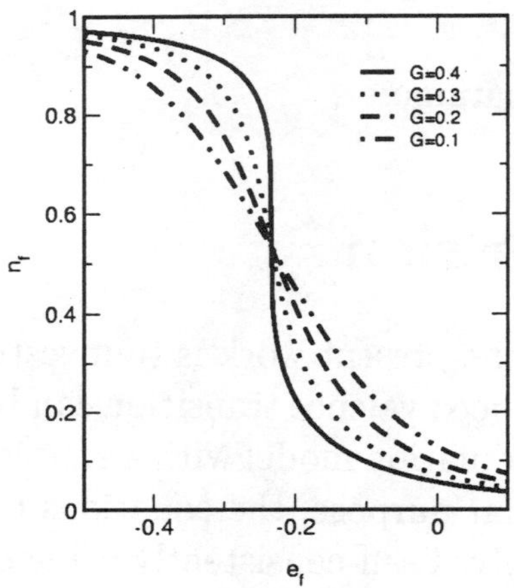

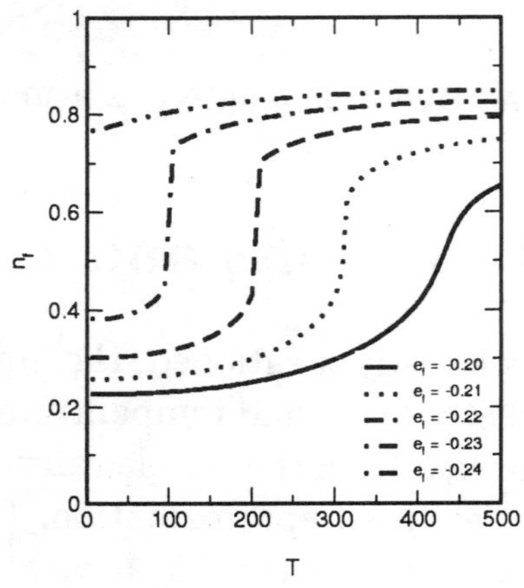

FIG.1: n_f as a function of e_f for various values of G at $T = 0$. $W = 1$, $V = 0.1$, $n = 1$ and $H = 0$.

FIG.2: n_f as a function of temperature for various values of e_f. $W = 1$, $V = 0.1$, $G = 0.4$, $n = 1$ and $H = 0$.

Guided by Fig.1, and restricting ourselves to the region of e_f where n_f changes rapidly, we investigated the temperature dependence of n_f for various close values of e_f in the absence of magnetic field. The results are striking and are displayed in Fig.2. For a given value of G and within a small region of e_f(for the some chemical compositions), n_f exhibits a discontinuous jump from a low to a high value at a critical temperature. On the other hand, for e_f values out side this narrow region, n_f either remains almost constant ($e_f = -0.20$) or varies smoothly($e_f = -0.24$). For e_f lying well-inside the band, the occupancy remains almost constant and on the other extreme location of e_f, n_f varies smoothly with increasing temperature. All these features of temperature induced valence transition were experimentally observed[1,2,4,11,12,13].

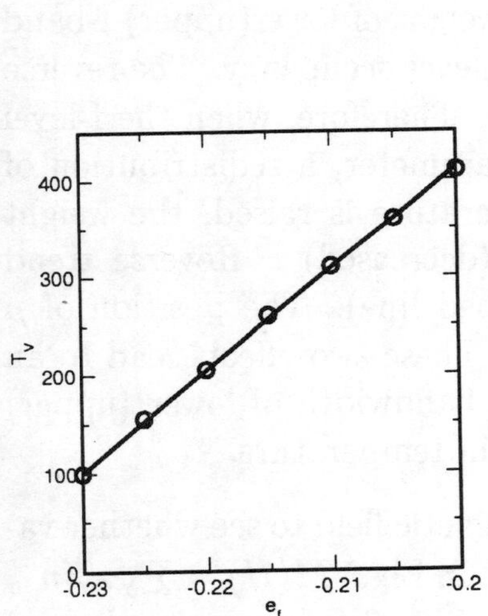

FIG.3: T_v as a function of e_f. Other parameters are same as in FIG.2

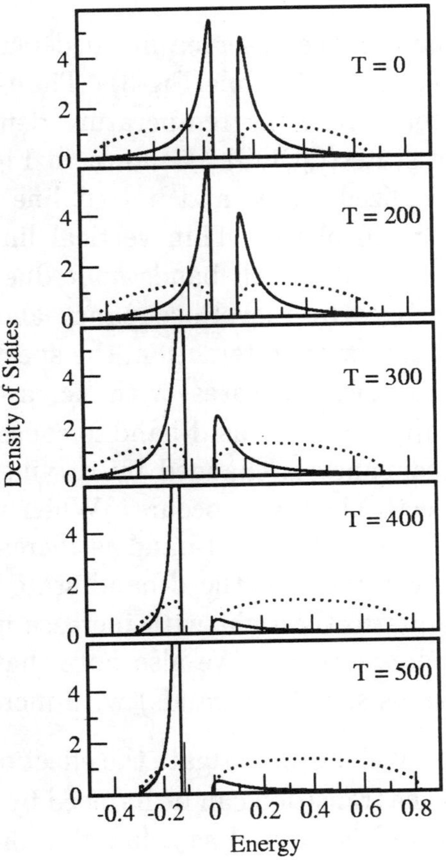

FIG.4: QDOS for various values of temperature for $e_f = -0.21$. Other parameters are same as in FIG.2.

For the cases where n_f changes sharply, one can define the valence transition temperature T_v as the T where $\partial n_f / \partial T$ has a peak. In Fig.3 T_v is plotted as a function of e_f and the dependence is found to be linear. The dependence of T_v on chemical composition as observed in Eu-compounds, $Eu(Pd_{1-x}Pt_x)Si_2^{15}$ and $EuNi_2(Si_{1-x}Ge_x)_2^{19}$, can be correlated with the change in e_f caused by the change in crystalline potential when the ligand composition is altered. As the crystalline structure remains same, it is expected that change in e_f would go linearly with x. In such a situation, namely, the observed linear dependence of T_v on x agrees well with the calculated one(Fig-3). The results shown so far can be understood from the temperature dependence of quasi-particle density of states(QDOS) as shown in Fig.4. In this figure, full line is for localized states and dotted line is for band-states in the paramagnetic phase. Thin vertical line shows the position of chemical potential. Both bands split due to hybridization which is renormalized by f-level occupancy and f-d correlation(eqns.(5-7)). Due to interstate interaction, the spectral weight of lower(upper) f-band increases (decreases) with increase in f-level occupancy. The reverse happens as far as d-band is concerned. Therefore, when the f-level occupancy is changed by varying a parameter, a redistribution of electronic states occurs. When temperature is raised, the weight of lower (upper) f-band is increased (decreased) . Reverse trend is observed for the d-band state (dashed line). The position of μ shifts to the right with increase in T. These two effects lead to an increase in n_f. We also note that the bandwidth of lower (upper) bands shrinks(expands) with increase in temperature.

We now investigate the effect of magnetic field to see whether valence transition can be induced by field. In Fig.5 $M(H) = \sum_\sigma z_\sigma (n_{f\sigma} + n_{d\sigma})$ is plotted as a function of H at $T = 0$ for various values of e_f. We see that for small fields, the magnetization is small and increases slowly with increasing field. In this region, the system is interpreted to be in the mixed valent state.As field goes beyond a characteristic field, $M(H)$ increases sharply and saturates. The system gets completely polarized and goes into the magnetic state. This behaviour closely resembles the observed behaviour of field dependence of magnetization in Eu-compounds[17,19,20]. The critical field where M sharply increases is found to be an increasing function of e_f. With increase of temperature lying between 0 and T_v, the jump in magnetization is diminished for a given e_f(chemical composition) and the critical field is reduced (Fig.6). Though not shown in figure, the sharp variation in magnetization altogether disappears for $T \gg T_v$ for a given e_f. In experiments also it was observed that for $T > T_v$ there is no field induced transition[14,15]. The results of the model calculations can again be understood by inspecting the QDOS which are plotted for different values of field at $T = 0$ and for different values of temperature at $H = 0.01$ in Fig.7 and Fig.8 respectively. In these figures, QDOS is plotted for spin-up in the positive half and for spin-down in the negative half of the frame.

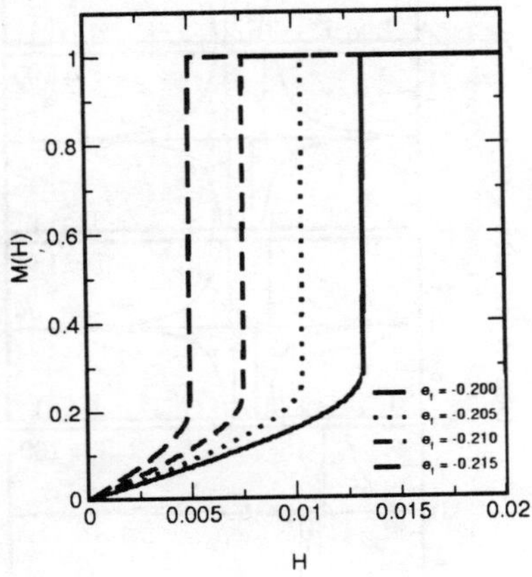

FIG.5: $M(H)$ as a function of magnetic field at $T = 0$ for various values of e_f. Other parameters are same as in FIG.2

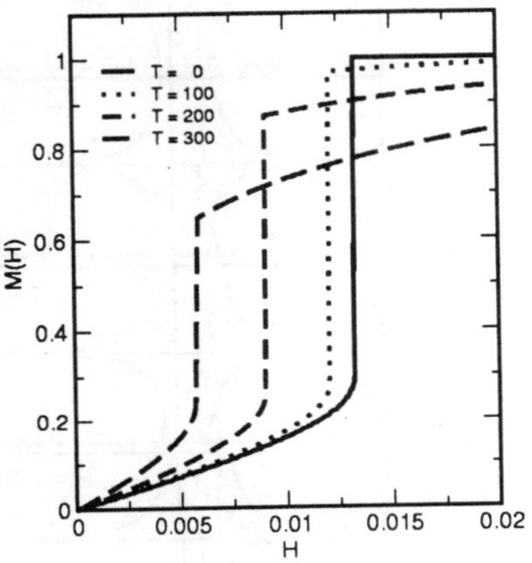

FIG.6: Field dependent magnetization as a function of magnetic field at $e_f = -0.2$ for various temperatures. Other parameters are same as in FIG.2

Again, here also the full line is for localized and dotted line for band states. Thin vertical line indicates the position of the chemical potential.When magnetic field is applied, due to the Zeeman splitting, the spin-up QDOS is pulled downwards and spin-down QDOS is pushed upwards. As a result $n_{f\uparrow}$ and $n_{d\uparrow}$ increase and correspondingly $n_{f\downarrow}$ and $n_{d\downarrow}$ decrease resulting in an increase in $M(H)$ with increasing H for small H. The Zeeman energy increases with the increase in magnetic field and when it exceeds the energy difference between the nonmagnetic and magnetic configurations, the system prefers to be in a magnetic configuration which leads to a sudden jump in $M(H)$ value. This can be seen from the sudden change in QDOS as one goes from $H = 0.013$ to $H = 0.015$ in Fig.7. Fixing the value of the field, the temperature-induced valence transition

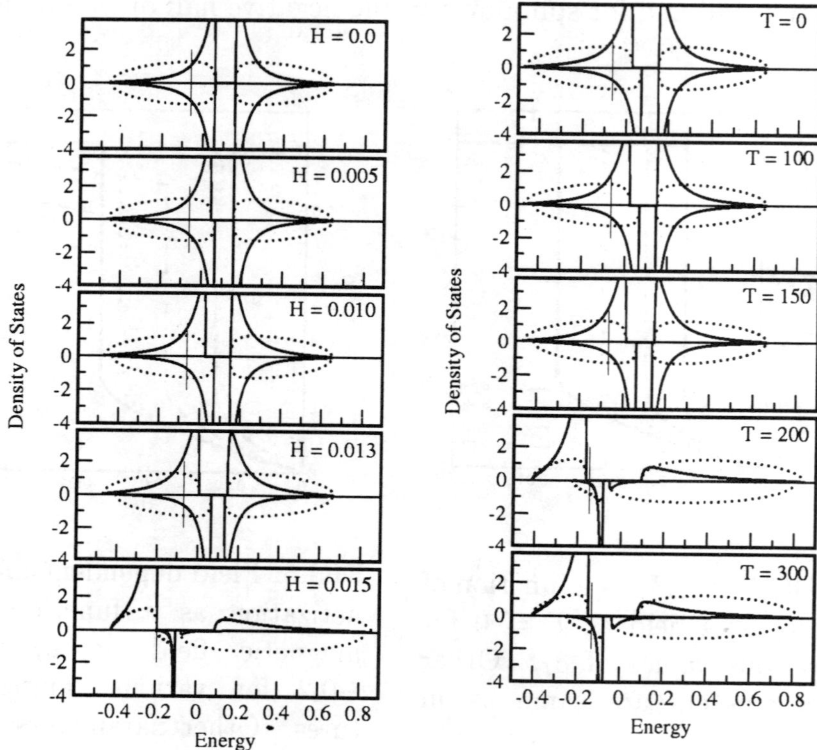

FIG.7: QDOS for various values of magnetic field at $T = 0$ and $e_f = -0.2$. Other parameters are same as in FIG.2.

FIG.8: QDOS for various values of temperature for $e_f = -0.2$ and $H = 0.01$. Other parameters are same as in FIG.2.

can be understood based on Fig.8. If one looks at QDOS for $T = 150$ and $T = 200$, there is a qualitative change in QDOS relative to the position of μ which leads to a jump in n_f. We have already seen from Fig.5 and Fig.6 respectively that the critical field H_c at which the metamagnetic transition takes place depends on both e_f and T. In Fig.9 is plotted the critical field at $T = 0$ as a function of e_f and the relation is linear. In experiments it was found that the critical field varies linearly with the chemical composition (Fig.4 in reference [19]). Finally, in Fig.10, we have plotted the reduced critical field

$H_c(T)/H_c(0)$ as a function of reduced temperature T/T_v for various values of e_f and all the points fall nearly on a smooth curve which is a sort of a universal

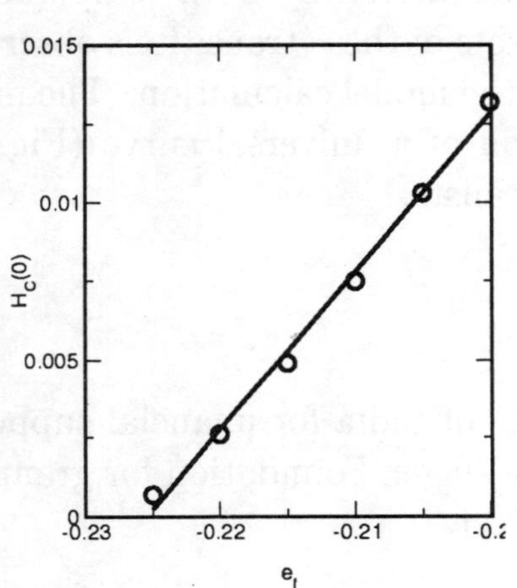

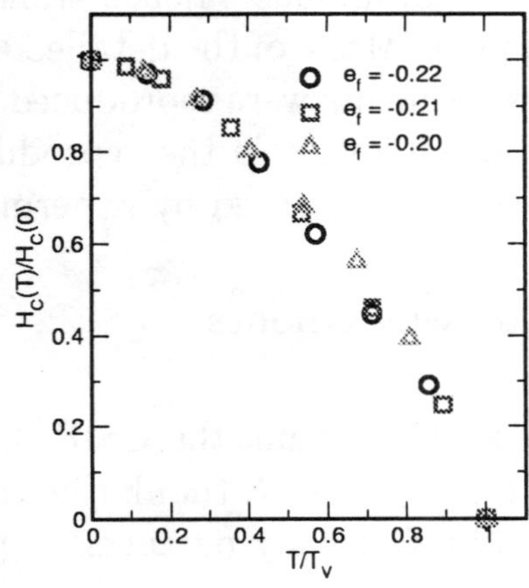

FIG.9: The critical field at $T = 0$ as a function of e_f. Other parameters are same as in FIG.2

FIG.10: Phase diagram in term of reduced critical field $H_c(T)/H_c(0)$ and reduced temperature T/T_v for $W = 1$, $V = 0.1$, $G = 0.4$ and $n = 1$

curve. It is very satisfying to find that this curve has a striking resemblance to the curves presented in reference [10] (Fig.4) and in reference [17] (Fig.1(b)) where the points corresponding to different compositions fall on a smooth curve. Thus, the simple calculation based on an electronic model for valence fluctuation, presented here, reproduces well all the qualitative features of field- and temperature-induced valence transition as observed in experiments.

4 Conclusions

We have made a model calculation using the EPAM with a focus on the the temperature- and field-induced valence transition. The Falicov-Kimball term in the model Hamiltonian has been treated in

HFA scheme suggested by Khomskii and Kocharjan and then the limit of infinite intrasite Coulomb repulsion is taken. The resulting coupled equations for the spin dependent average occupations of f-level have been self-consistently solved with total number of electrons per lattice site fixed at one. We have found both continuous and discontinuous valence transitions driven by both temperature and field. Many of the detailed aspects of these transitions observed in experiments were reproduced by the model calculation. The most significant result is the reproduction of a universal curve (Fig.10) which was presented by experimentalists.

Acknowledgments

The authors thank the CSIR, Govt. of India for financial support. A. R. and G. G. R. thank the Volks Wagen Foundation for granting an India-Germany partnership project.

References

[1] I. Felner and I. Nowik, Phys. Rev.B **33** 617 (1986).

[2] I. Felner, I. Nowik, D. Vaknin, U. Potzel, J. Moser, G. M. Kalvius, Wortmann G., G. Schmeister, G. Hilscher, E. Gratz, C. Schmitzer, N. Pillmayr, K. G. Prasad and H. de Ward, Phys. Rev. **B 35** 6956 (1987).

[3] K. Sugiyama, F. Iga, T. Kasaya and M. Date, Jour. Phys. Soc. Japan **57** 3946 (1988).

[4] K. Yoshimura, T. Nitta, M. Mekata, T. Shimizu, T. Sakakibara, T. Goto and G. Kido, Phys. Rev. Lett. **60** 851 (1988).

[5] K. Kojima, H. Hayashi, A. Minami, Y. Kasamatsu and T. Hihara, J. Magn. Magn. Mater. **81** 267 (1989).

[6] T. Shimizu, Y. Yoshimura, T. Nitta, T. Sakakibara, T. Goto and M. Mekata, J. Phys. Soc. Japan **57** 405 (1988).

[7] C. Rossel, K. N. Yang, M. B. Maple, Z. Fisk, E. Zimgiebl and J. D. Thompson, Phys. Rev. **B 35** 1914 (1987).

[8] N. Pillmayr, E. Bauer and Y. Yoshimura, J. Magn. Magn. Mater. **104-107** 639 (1992).

[9] H. Aruga Katori, T. Goto and K. Yoshimura, Physica **B 201** 159 (1994).

[10] H. Aruga Katori, T. Goto and K. Yoshimura, J. Magn.Magn. Mater. **140-144** 1245 (1995).

[11] E. V. Sampathkumaran, L. C. Gupta, R. Vijayaraghavan, K. V. Gopalakrishnan, R. G. Pillay and H. G. Devare, J. Phys. C **14** L237 (1981).

[12] E. Kemly, M. Croft, V. Murgai, L. C. Gupta, C. Godart, R. D. Parks and C. U. Segre, J. Magn. Magn. Mater. **47&48** 403 (1985).

[13] G. Wortmann, K. H. Frank, E. V. Sampathkumaran, B. Perscheid, G. Schmeister and G. Kaindl, J. Magn. Magn. Mater. **49** 325 (1985).

[14] . H. Wada, A. Mitsuda, M. Shiga, H. Arug Katori and T. Goto, J. Phys. Soc.Japan **65** 3471 (1996)

[15] A. Mitsuda, H. Wada, M. Shiga, H. Aruga Katori and T. Goto, Phys. Rev.B **55** 12474 (1997).

[16] A. Mitsuda, H. Wada, M. Shiga and T. Tanaka, J. Phys.: Condens. Matter **12** 5287 (2000)

[17] A. Mitsuda, H. Wada, M. Shiga, H. Arug Katori, H. Mitamura and T. Goto, J. Magn. Magn. Mater. **196-197** 883 (1996).

[18] G. Wortmann, I. Nowik, B. Perschied and G. Kaindl 1991 Phys. Rev. **B 43** 5261(1991).

[19] H. Wada, A. Nakamura, A. Mitsuda, M. Shiga, T. Tanaka, H. Mitamura and T. Goto, J. Phys.: Condens. Matter **9** 7913 (1997).

[20] H. Wada, A. Nakamura, A. Mitsuda, M. Shiga, H. Mitamura and T. Goto, J.Magn. Magn. Mater. **177-181** 363 (1998).

[21] H. -J. Hesse, R. Lübbers, M. Winzenick, H. W. Neuling and G. Wortmann, J. of Alloys and Compounds **246** 220 (1997).

[22] D. Khomskii and A. N. Kocharjan, Solid St. Commun. **18** 985 (1976).

[23] L.V. Keldysh and Yu. V. Kopaev, Fiz. Tverdovo Tela **18** 1686 (1964).

[24] L.M. Falicov and J.C. Kimball, Phys. Rev. Lett. **22**, 997 (1969).

[25] H. J. Leder, solid state Commun. **27**, 579 (1978).

[26] A.P.G. Kutty J. Phys. Chem. Solids **45** 121(1984).

[27] C.E.T Goncalves da Silva and L.M. Falicov, solid State Commun. **17**, 1521 (1975).

[28] Ishwar Singh, A.K. Ahuja and S.K. Joshi Solid state Commun. **34**, 65 (1980).

[29] G. Gangadhar Reddy and A. Ramakanth, J. Phys. Chem Solids **51** 515 (1990).

[30] C.M. Varma and Y. Yafet, Phys. Rev. **B 13**, 2950 (1975).

[31] G. D. Mahan, Many Particle Physics (Plenum, New York, 1981)

[32] R. J. Jellito, J. Phys. Chem Solids **30** 609 (1969).

Physics of Solids, Nuclei and Particles
Editor: R. Sahu

Recent Trend in Electronic Science

J. Majhi

Professor of Physics (Retd.), Indian Institute of Technology Madras
Chennai 600 036, India

I. Introduction

Science in general has progressed over the years right from BC to present century. The earlier developments were based on purely observations of natural phenomena and accumulation of facts. Systematic theoretical and experimental investigations as a means of scientific enquiry began during 16th to 18th century by Kepler, Galileo, Newton and many others. This period saw the growth of mainly mechanics, optics and heat and thermodynamics. Developments in electricity started in 19th century with the pioneering work of Michael Faraday, Lord Kelvin, Maxwell, 1.1. Thomson and others. The naturally available materials like Ag_2S, Se, PbS, CuO etc have been used as the electronic materials in the earlier solid state devices. Since then electronic science and technology has grown into greater heights. The 20th century can be marked as the century of electronics since it has witnessed the development of vacuum tubes, invention of transistor and large scale integrated circuits (VLSI). To explain the characteristics of new materials and devices, electronic theory has gone through a number of changes from classical ideas to quantum concepts.

II. Progress in the Electronic Materials

Silver sulphide Ag_2S seems to be the first electronic material identified by Faraday in 1833 as a photo-conductor. Later Braun (1874) and 1.C.Bose (1904) used lead sulphide PbS for rectification. However, due to inadequate purity and non-crystallinity of the naturally available sulphides and oxides, these materials could not produce reliable devices and hence the growth of solid state devices apparently slowed down. On the other hand, with the advancement in vacuum technology, vacuum tubes (valves) were developed and dominated the field of electronics during 1920s to 1950s. The real breakthrough in electronic materials came with the synthesis and crystal growth of high purity germanium Ge and silicon Si in 1940s. Almost 90including VLSI are fabricated on silicon wafers. However, silicon being a indirect band gap material and of smaller band-gap (1.1 e V) is not suitable for optoelectronic devices like LEDs and LASER diodes. Direct band gap materials GaAs (1.42 eV) and its alloys are extensively used in optoelectronic and quantum well devices. Unlike silicon, GaAs surface cannot be easily oxidized or passivated which are essential for the fabrication of integrated circuits.

III. Progress in Solid State Devices

As mentioned earlier, the first solid state rectifier (PbS) was discovered by Braun in 1874 much before the vacuum tube diode was developed (1920s). But due to the material problem the initial growth of solid state devices was slow, Development of vacuum tube devices almost shadowed the the development of solid state devices until 1940s. Again the availability of pure Ge and Si and the invention of transistor in 1947 by Bardeen, Brattain and Shokley revived the progress in semiconductor devices. Solid state devices have several advantages over the vacuum tubes, not only in terms of size but also in reliability and stability. Advancement in solid state devices was possible mainly due to extensive research and development in material science allover the world. The silicon planar technology gave birth to the integrated circuits ICs or microelectronics. Along With the progress in microelectronics there was also a rapid development of optoelectronic devices like optical detectors (LDR,photo diode,photo transistor etc), solar cells, optical sources (LEDs,Laser diodes). Miniaturization of devices has reached almost the physical limitations leading to nanostructure devices and quantum dots. The nano-technology has progressed well in recent years while the nanoscience seems to be lagging behind since there are several challenging problems in nanostructure devices yet to be solved theoretically.

IV. Progress in Electronic Science

Electricity received great attention during 18th and 19th century. The pioneers in this period are Benjamin Franklin, Henry Cavendish, Charles Coulomb, Ampere, Volta, Faraday, Maxwell, Lorentz, Helmholtz, Hertz and many others. On the theoretical side Biot-Savart law of magnetic induction due to a current element, Maxwell's theory of electromagnetic field, Faraday's laws of electrolysis etc are some of the important contributions to the field of electricity. The earlier fluid theory of electrical conduction was discarded after the discovery of electron in 1897 by J.J. Thomson. Evolution of quantum mechanics and the subsequent development of free electron theory and band theory of solids by 1930s helped to explain the phenomenon of electronic conduction in metals and semiconductors. The concept of energy bands, energy gap, effective mass and Fermi level became handy and helped to understand the optical and electronic properties of materials and devices. The famous theories of rectification by Schottky (1939) on metal-semiconductor contact and Shockley (1949) on P-N junction became the foundation to explain the basic characteristics of most of the solid state devices available today. The Maxwell's equations, Boltzmann transport equations, continuity equations and Schrodinger equations are frequently used to explain the properties of semiconductor materials and devices. These equations are suitably modified for the disordered materials. The classical Boltzmann transport equation has been extended to ballistic and quantum transport of charge carriers and Monte Carlo method is introduced for numerical solution of Boltzmann equation. Scattering theory is extensively applied to explain the the mobility of charge carries in semiconductors affected by phonon scattering, impurity scattering, grain boundary and surface scattering. The theory of tunneling has been used to explain the characteristics of Zener diodes, tunnel diodes and backward diodes and the recent resonant tunneling devices. High frequency oscillation in Gunn diode and tunnel diode has been explained on the basis of negative differential resistance. The progress in semiconductor technology encouraged new theoretical developments in the mechanism of crystal growth (bulk and epitaxial), diffusion of impurities (thermal and ion-implantation), oxidation (thermal and plasma) in silicon, lithography (optical and electron beam) etc. Transport of charge carriers in short channel FETs, in VLSI and nanostructure devices are being investigated. The theory of P-N junction (homojunction) has been extended to heterojunction devices like LEDs, Laser diodes, quantum well devices, quantum dots etc.

V. Some New Devices

V.I. Super conducting Device: With the development of high temperature superconductivity there is a hope for realizing very high speed and low noise devices using the principle of Josephson junction. Besides the high cost there are difficulties in interfacing and degradation in these devices. Some R & D work is going on in Japan.and USA.

V.II. Resonant Tunneling Device: Two parallel hetero-junction barriers separated by a small distance (50 A) constitute a resonant tunneling device. The I-V characteristic of such a device gives rise to negative differential resistance and hence can produce high frequency oscillations. Theoretically a self consistent solution of the Schrodinger equation including the Maxwell's equation is being tried.

V.III. Nanostructure Devices: The IC technology has gone through revolutionary changes over a short span of time from small scale integration to large scale integration (LSI) to very large scale integration(VLSI) or ultra large scale integration (ULSI) . Thus the device size (channel length) has shrunk from micrometers to nanometers. The characteristics of nanoscale devices are completely different from the conventional devices and needs new theoretical interpretations including the Uncertain Principle.

V.IV. Organic Electronics: Organic materials like polyacetylene have shown promises to be used in electronic devices due to their high electron mobility along the chain. The electrical, optical and chemical properties of these materials can be changed by oxidation or reduction, and by incorporating different chemical groups.

V.V. Bio-electronics: Information processing in living systems is known to be of electrical in nature. However, unlike in electrical circuits the charge carriers are ions rather than electrons and hence the transmission signal is slow and weak.

VI. Conclusion

All round growth in science and technology of semiconductor devices took place during the twentieth century. Many new devices in microelectronics and opto-electronics are invented and the corresponding theories developed. Solid state electronics being interdisciplinary in nature, could encourage the growth of almost all branches of science and engineering. Semiconductor devices now find a very wide range of applications ranging from home appliances to sophisticated instruments. Miniaturization of devices has reached almost the physical limitations. Day by day the size of the computer is becoming smaller with higher memory capacity. The electronic industry is ever growing and is bound to give us many more new devices in the coming years.

Physics of Solids, Nuclei and Particles
Editor: R. Sahu

Reduction of Quantum Mechanical Noises in Photonic and Atomic Systems

N. Nayak

S.N. Bose National Centre for Basic Sciences, Block-JD, Sector-3, Salt Lake City
Kolkata 700 098, India

I. Introduction

Since the advent of quantum mechanics, there have been persistent endevours to obtain unconventional fields of electromagnetic radiation (photonic field) as well as atomic states having properties smarter than known ones. Unconventional properties mean that they owe their allegiance to quantum mechanics. In other words, they do not have any classical counterpart. Essentially, what is special in these systems is that the disturbances or noises are at their minimum. Among such endevours, laser (acronym for light wave amplification by stimulated emission of radiation) is one of the most significant achievements in the last century. The electromagnetic radiation coming out of a laser device does have noises reduced to their minimum as per the rules of quantum mechanics. Such fields are in a state which is the so-called *minimum uncertainty state* or the MUS since this state satisfies the equality sign in the uncertainty relation. This was defined by Schrödinger in 1926 [1]. In the present day terminology, they are known as coherent states. So, to-day we can have radiation fields (from a laser) in a coherent state as well as the atomic or molecular states (prepared by a laser) can have properties of coherent states.

It is now a well established fact that lasers have brought novel and extremely useful changes (not possible with out application of a laser) in science and technology. However, as mentioned above, laser radiations are not free of noise. Indeed, they have noise reduced to a minimum in that for a MUS. Hence, their applications are difficult at places where the minimum quantum mechanical noises are even crucial. One such example is its application in the detection of waves of the gravitational field. This prompts us to look for ways to obtain electromagnetic radiation which has noise less than that for a MUS and at the same time does not violate the uncertainty principle. We shall see in the following sections that application of lasers to certain nonlinear optical mediums (where the electric polarization is not proportional to the optical field) does produce such radiation field which does not have a classical counterpart. Electromagnetic field with these properties are called squeezed radiation.

The next question that comes to mind is whether we can have similar states representing atomic levels. We know that any two levels of an atom or molecule (which we show below also) is equivalent to a spin-half particle. If spin-up represents the upper of the two levels, then spin-down represents the lower level. An ensemble of atoms where each atom is represented by its two levels, can be represented by a spin-half system. When this system is in a MUS, then the state representing the system is known as a spin coherent state or a atomic coherent state [3]. When the noise is reduced further down the minimum noise in the spin coherent states, then the state is said to be spin squeezed. These states have applications in the ultra high precision technology, such as atomic clock, the world's ultimate clock having accuracy upto fourteenth place after the decimal point. In other words, it will have an error of 1 part in 10^{14}.

In the following, we address to further details of squeezing of the noise. Towards this end, we derive the quantization of the electromagnetic field and, also, show the equivalence between a two-level atom and a spin-half particle for a self-contained reading.

II. Quantization of the Electromagnetic Field

This has been dealt in many text books, but, there the quantization is done for an unbounded region using vector potential. What we need for the present purpose is the quantized version of the radiation field in a cavity, that is, a scalar field. It is unnecessary to treat the problem using vector potential and, so, we treat the quantization of the electromagnetic field using electric and magnetic field. This gives the quantized version appropriate for typical problems in quantum optics, in general. The Maxwell's equations in a source-free space read as

$$\nabla \cdot \mathbf{D} = 0, \tag{1}$$

$$\nabla \cdot \mathbf{B} = 0 \tag{2}$$

with $\mathbf{D} = \epsilon_0 \mathbf{E}$ and $\mathbf{B} = \mu_0 \mathbf{H}$,

$$\nabla \times \mathbf{H} = \frac{\partial \mathbf{D}}{\partial t}, \tag{3}$$

and

$$\nabla \times \mathbf{E} = -\frac{\partial \mathbf{B}}{\partial t}. \tag{4}$$

The coupled equations can be cast in the form

$$\nabla^2 \mathbf{E} - \frac{1}{c^2}\frac{\partial^2 \mathbf{E}}{\partial t^2} = 0 \tag{5}$$

governing the electric field only and

$$\nabla^2 \mathbf{H} - \frac{1}{c^2}\frac{\partial^2 \mathbf{H}}{\partial t^2} = 0 \tag{6}$$

for the magnetic component of the electromagnetic field. We shall quantize the electric field only. In the interaction between electromagnetic field and atom or molecule, both the electric and the magnetic components interact with the charged particles (electrons).

They are know as dipole or, in general, multipole interactions, The electric dipole interaction is 10^{16} times stronger than the magnetic component of the radiation field. For this reason the electric field is known as the optical field. Hence, the quantization of the electric field will suffice the present purpose.

We take the electric field to be linearly polarized and oscillating in a cavity of length L. Then we can expand the electric field in the normal modes $\mathcal{U}$ of the cavity

$$E_x(z,t) = \sum_i A_i \mathcal{U}_i(z,t) \tag{7}$$

with

$$\mathcal{U}_i(z,t) = q_i(t)\sin(k_i z) \tag{8}$$

where x is the direction of polarization and z is the axis of the cavity. The A_i are given by

$$A_i = \left[\frac{2\Omega_i^2 m_i}{V\epsilon_0}\right]^{1/2}$$

where Ω_i is the eigen frequency of the mode i and V is the volume of the cavity. The m_i is a constant with the dimension of mass which has been included only to establish the analogy between the dynamics of the mode with that of a simple harmonic oscillator having mass m_i and a Cartesian coordinate q_i. The non-vanishing component of the magnetic field H_y can be obtained from Eq. (3) and is given by

$$H_y(z,t) = \sum_i A_i\left[\frac{\dot{q}\epsilon_0}{k_i}\right]\cos(k_i z). \tag{9}$$

The classical Hamiltonian for the field

$$\mathcal{H} = \frac{1}{2}\int_V d\tau(\epsilon_0 E_x^2 + \mu_0 H_y^2) \tag{10}$$

takes the form, in terms of q_i and $\dot{q}_i$,

$$\mathcal{H} = \frac{1}{2}\sum_i(m_i\Omega_i^2 q_i^2 + m_i p_i^2) \tag{11}$$

with $p_i = m_i \dot{q}_i$. This shows that each mode of the electric field is equivalent to a mechanical harmonic oscillator. The dynamics can be quantized if we identify q_i and p_i as operators obeying

$$[q_i, p_j] = i\hbar \delta_{ij} \tag{12}$$

and

$$[q_i, q_j] = [p_i, p_j] = 0. \tag{13}$$

We now make a canonical transformation to operators a_i and $a_i^\dagger$ given by

$$a_i e^{-i\Omega_i t} = \frac{1}{\sqrt{2m_i \hbar \Omega_i}} (m_i \Omega_i q_i + i p_i) \tag{14}$$

and

$$a_i^\dagger e^{i\Omega_i t} = \frac{1}{\sqrt{2m_i \hbar \Omega_i}} (m_i \Omega_i q_i - i p_i) \tag{15}$$

respectively, by which $\mathcal{H}$ takes the form

$$\mathcal{H} = \hbar \sum_i \Omega_i (a_i^\dagger a_i + 1/2). \tag{16}$$

From Eqs. (12)-(15), it follows that a_i and $a_i^\dagger$ obey the commutation relation

$$[a_i, a_j^\dagger] = \delta_{ij}. \tag{17}$$

The a_i and $a_i^\dagger$ are called the annihilation and creation operators respectively. They operate as

$$a_i |n_i\rangle = \sqrt{n_i} |n_i - 1\rangle,$$

$$a_i^\dagger |n_i\rangle = \sqrt{n_i + 1} |n_i + 1\rangle,$$

and the number operation $a_i^\dagger a_i$ is defined by

$$a_i^\dagger a_i |n_i\rangle = n_i |n_i\rangle.$$

In terms of a_i and $a_i^\dagger$, the electric field takes the form

$$E_x(z, t) = \frac{1}{2} \sum_i \mathcal{E}_i (a_i e^{-i\Omega_i t} + a_i^\dagger e^{i\Omega_i t}) \sin(k_i z) \tag{18}$$

where $\mathcal{E}_i = 2\left[\frac{\hbar\Omega_i}{V\epsilon_0}\right]^{1/2}$ has the dimension of electric field. We immediately see that $\langle E \rangle = 0$, but,

$$\langle E_x(z)^2 \rangle = \frac{1}{2}\mathcal{E}_i^2\left(n_i + \frac{1}{2}\right)\sin^2 k_i z \qquad (19)$$

for the cavity mode i in a Fock state $|n_i\rangle$ containing exactly n_i photons. This does not go against quantum mechanics since it is not $\langle E \rangle$ but the intensity of the radiation field $I \propto \langle E_x(z)^2 \rangle$ is an observable. However, such cases do not happen for all kinds of radiation field. One such example is the field in a coherent state in which $\langle E \rangle \neq 0$. With this information we now describe what is known as squeezed radiation.

III. Squeezing of the Electromagnetic Field

We start with the minimum uncertainty state (MUS) as defined by Schrödinger [1]. It is an eigen state of an non-Hermitian operator and is given by, in general,

$$[q + i\xi p]|MUS\rangle = [\langle q \rangle + i\xi\langle p \rangle]|MUS\rangle \qquad (20)$$

where ξ is a real constant. The operators q and p are the canonical coordinate and momentum respectively. Note that this is a general equation and it covers the radiation field, defined in the last section, as well. If we set $\xi = 1/m\Omega$ and, with little algebra, we can show that Eq. (20) reduces to

$$a|MUS\rangle = \alpha|MUS\rangle \qquad (21)$$

where a is the annihilation operator given by Eq. (14) with $t = 0$. Since $|MUS\rangle$ is an eigenstate of a non-Hermitian operator a, α is, in general, complex and is given by

$$\alpha = \sqrt{\frac{m\Omega}{2\hbar}}\langle q \rangle + \frac{i}{\sqrt{2m\Omega\hbar}}\langle p \rangle. \qquad (22)$$

In the present day terminology they are known as coherent states [2] and are usually represented by $|\alpha\rangle$.

Let us now get back to the Eq. (18) and re-write as

$$E_x(t) = \frac{1}{2}(ae^{-i\Omega t} + a^\dagger e^{i\Omega t}) \tag{23}$$

where we have set the number of modes $i = 1$. Also, we have omitted $\mathcal{E}$ and the mode structure $\sin(kz)$ which can be taken care of when the need arises. The Eq. (23) carries all the features of quantum mechanics. We can write Eq. (23) as

$$E_x(t) = a_c \cos(\Omega t) + a_s \sin(\Omega t) \tag{24}$$

where $a_c = (a + a^\dagger)/2$ and $a_s = (a - a^\dagger)/2i$ are the amplitudes of the cosine and sine components of the electric field. With the help of the Eq. (17), we see that the quadrature operators satisfy the commutation relation

$$[a_c, a_s] = i/2. \tag{25}$$

Since they do not commute, we know from the uncertainty relation that their root-mean-square deviations must satisfy

$$\Delta a_c \Delta a_s \geq \frac{1}{4}. \tag{26}$$

If the optical field represented by the Eqs. (23) and (24) is in a MUS or a coherent state, then it must further satisfy

$$\Delta a_c = \Delta a_s = \frac{1}{2} \tag{27}$$

with their product satisfying the equality sign in the Eq. (26) as per the definition for the MUS in Eq. (21). The associated noise can be obtained after multiplying to the right side of Eq. (27) by the appropriate parameters representing the physics of the problem. Note that the right side of the Eq. (26) is a c-number quantity. If it would have been an operator (observable), then we had to find out the minimum eigenvalue of that operator. This is the case with spins (Fermions) which we will see in the following sections.

The radiation from a well stabilized laser has the property in Eq. (27), that is, the laser radiation is indeed in a coherent state. However, as mentioned in the introduction above, it just happens that the minimum noise given by Eq. (27) is still a limiting factor in certain experiments such as detection of gravity waves. Thus the question arises if we can reduce Δa_c or Δa_s below their minimum values 1/2 which at the same time satisfy the fundamental quantum mechanical relation, the uncertainty relation in Eq. (26). Such a possibility is there. That is, we can have

$$\Delta a_c < 1/2 \tag{28}$$

at the expense of Δa_s or the vice versa such that Eq. (26) is satisfied. Radiation having this property is called squeezed radiation. Such a nomenclature is due to the fact that noise in one component is taken out (squeezed out) which simultaneously increases the noise in the complimentary component.

The generation of squeezed radiation has been experimentally demonstrated [4]. For further information, the reader may consult the special issue in the Journal of the Optical Society of America in Ref. [5]. In the generation of squeezed radiation, a particular feature of the Hamiltonian representing the atom-field interaction is that it should be at least second order in the radiation field operators a and $a^\dagger$. Such a Hamiltonian describes what is known as two-photon process. To understand this process, let us consider an ensemble of atoms/molecules where each of them are described by three levels, namely, upper level $|a\rangle$, intermediate level $|i\rangle$ and lower level $|b\rangle$. We apply radiation from a laser to this ensemble such that the field frequency ν is not in resonance with the $|b\rangle \leftrightarrow |i\rangle$ and $|i\rangle \leftrightarrow |a\rangle$ transitions. But the frequency between $|a\rangle$ and $|b\rangle$ is equal to the twice of the laser radiation frequency, 2ν. This process can be described by a Hamiltonian

$$\mathcal{H} = g_1(s_1^+ a + s_1^- a^\dagger) + g_2(s_2^+ a + s_2^- a^\dagger) \tag{29}$$

where the s are the atomic operators which we describe clearly later. g_1 and g_2 are the atom-field interaction strengths responsible for the transition $|b\rangle \leftrightarrow |i\rangle$ and $|i\rangle \leftrightarrow |a\rangle$ respectively. We can follow the standard perturbation theory and go upto the second order proportional to $g_1 g_2$ to eliminate the non-resonant level $|i\rangle$. Then, Eq. (29) reduces to the effective Hamiltonian

$$\mathcal{H}_{eff} = g(s^+ a^2 + s^- a^{\dagger 2}) \tag{30}$$

where $s^+ = s_2^+ s_1^+$ and $s^- = s_2^- s_1^-$. g is called the effective coupling constant for the two-photon transition. We see in Eq. (30) that the Hamiltonian is quadratic in field operators and, also, is capable of squeezing the radiation field. There is another process in which the Hamiltonian in Eq. (30) represent one of the interactions there and this process is capable of squeezing the radiation field. This process is called the optical parametric process in which the atom absorbs a pump photon (applied laser field) and makes a transition from the lower level $|b\rangle$ to the upper level $|a\rangle$. But, the downward process involves emission of two photons called the signal and the idler via the intermediate level $|i\rangle$. This process has been utilized for the experimental demonstration of squeezing the radiation field [4]. Likewise, there are quite a few processes that have been discussed in the literature [5] showing their capability in squeezing the interacting radiation field out of which a few have been utilized for its experimental demonstration.

Thus we find that it is possible to squeeze the noise in the radiation field below that for a MUS. This prompts us to find out if one can squeeze the noises associated with atomic/molecular transitions. Before we attempt in this direction, it is worthwhile to discuss a striking analogy between spin-half particles (Fermions) and a two-level atom or a molecule.

IV. Analogy between a Fermion and a Two-level Atom

Let us consider electronic transitions in an atom or a molecule. We focus on the two atomic levels whose frequency difference is equal to or nearly equal to the frequency of the interacting radiation field. The other energy levels are not important for consideration here since their transition frequencies are far away from resonance. In the language of quantum optics, this is known as a two-level atom. So, we represent the atom/molecule by these two energy levels only, the upper level $|a\rangle$ and the lower level $|b\rangle$. An ensemble of such atoms or molecules are as a two-level system. We now show that the two-level atom is analogous to a fermionic particle having spin 1/2. To this end let us consider the operators

$$s^+ = |a\rangle\langle b|, \tag{31}$$

$$s^- = (s^+)^\dagger = |b\rangle\langle a|, \tag{32}$$

and

$$s_z = \frac{1}{2}[|a\rangle\langle a| - |b\rangle\langle b|]. \tag{33}$$

We find that $s^+|b\rangle = |a\rangle$, that is, the operator s^+ takes the atom from lower level to the upper level. Also, $s^-|a\rangle = |b\rangle$, $s_z|a\rangle = \frac{1}{2}|a\rangle$, and $s_z|b\rangle = -\frac{1}{2}|b\rangle$. In the units of frequency of transition ω, s_z gives the population of the upper and the lower levels with the zero of the energy ladder being fixed at the middle such that the energy of $|a\rangle$ and $|b\rangle$ are $\frac{1}{2}\hbar\omega$ and $-\frac{1}{2}\hbar\omega$ respectively. We further define

$$s_x = \frac{s^+ + s^-}{2} \tag{34}$$

and

$$s_x = \frac{s^+ - s^-}{2i}. \tag{35}$$

With the help of the Eqs. (31)-(33) we find that the s_x, s_y, and s_z do not commute with one another. In fact they satisfy

$$[s_x, s_y] = i s_z \tag{36}$$

The other two commutation relations are obtained by cyclic permutation of x, y, and z in Eq. (36). We see from Eqs. (31)-(36) that this algebra is also followed by spin-half operators. In deed, the s_x and s_y give the x- and y-components of the atomic polarization respectively. But the z-component, the s_z, gives the population of the atomic levels unlike the z-component of the spin operator. $\langle s_z \rangle = +1/2$ when the atom is in the upper state and, when $\langle s_z \rangle = -1/2$, then the atom is in the lower state. So, the range of values it can take is given by $-\frac{1}{2} \leq \langle s_z \rangle \leq \frac{1}{2}$. For this reason, the atomic/molecular operators defined in Eqs. (31)-(36) are known as pseudo-spin operators.

Thus, a two-level system of N atoms with operators

$$S_x = \sum_N s_x, \quad S_y = \sum_N s_y, \quad and \quad S_z = \sum_N s_z \qquad (37)$$

are equivalent to a system with total spin $N/2$ such that we have the condition

$$-\frac{N}{2} \leq \langle S_z \rangle \leq \frac{N}{2}. \qquad (38)$$

V. Spin Squeezing

It is easy to see that the operators in the Eq. (37) obey similar commutation relations as in Eq. (36) which suggests that

$$\langle \Delta S_x \rangle \langle \Delta S_y \rangle \geq \frac{1}{2} |\langle S_z \rangle| \qquad (39)$$

where $\langle \Delta S_x \rangle = \sqrt{\langle S_x^2 \rangle - \langle S_x \rangle^2}$ and a similar expression for $\langle \Delta S_y \rangle$. We see in Eq. (39) that the uncertainty product is not a minimum when the equality sign in taken.This is the case with the electromagnetic field which is due to the fact that the right hand side of the uncertainty product satisfied by it is a c-number quantity [see Eq. (26)]. Whereas in the atomic case, it is the observable $\langle S_z \rangle$ which appears on the right side of its uncertainty relation. So, we have to find out the minimum this observable can assume in order to find out a MUS for the two-level system. From the range of values that $\langle S_z \rangle$ can take [Eq. (38)], it is evident that

$$\langle S_z \rangle_{min} = -\frac{N}{2} \qquad (40)$$

Thus, an atomic MUS or atomic coherent state [3] must satisfy

$$\langle \Delta S_x \rangle \langle \Delta S_y \rangle = \frac{N}{4} \tag{41}$$

with equal uncertainty in the x- and the y-component

$$\langle \Delta S_x \rangle = \langle \Delta S_y \rangle = \frac{\sqrt{N}}{2}. \tag{42}$$

This is rather not new to us. The Wigner or the angular momentum states, that we know, can have this property. For example, consider the Wigner state $|j, m\rangle$ describing the angular momentum of magnitude $\mathbf{J} \equiv (J_x, J_y, J_z)$. We have $J^2|j, m\rangle = j(j + 1)|j, m\rangle$ and $J_z|j, m\rangle = m|j, m\rangle$. We also have $[J_x, J_y] = iJ_z$ and, hence,

$$\langle \Delta J_x \rangle \langle \Delta J_y \rangle \geq \frac{1}{2}|\langle J_z \rangle|. \tag{43}$$

We also know that $\langle J_x \rangle = \langle J_y \rangle = 0$, but, the root-mean-square deviations $\langle \Delta J_{x,y} \rangle = \sqrt{\langle J^2_{x,y} \rangle - \langle J_{x,y} \rangle^2}$ have equal magnitudes

$$\langle \Delta J_x \rangle = \langle \Delta J_y \rangle = \frac{1}{\sqrt{2}} \sqrt{j(j + 1) - m^2}. \tag{44}$$

We find that, for $m = \pm j$, the variances reduce to their minimum $\langle \Delta J_x \rangle = \langle \Delta J_y \rangle = \sqrt{j/2}$. Thus, as per the definitions in the Eqs. (41)-(42) the angular momentum states $|j, \pm j\rangle$ are indeed minimum uncertainty states.

Thus, in the vectorial representation of an atomic coherent states, the vector is along the z-axis (say) and the variances in the x- and y-components in the plane normal to the z-axis take a minimum value. Let us now consider a two-level system of N atoms represented by the spin vector $\mathbf{S} \equiv (S_x, S_y, S_z)$ with total spin $S = N/2$. We represent its direction by the polar angles θ and ϕ. The θ gives the angle between the vector $\mathbf{S}$ and the vertical z-axis. The ϕ represents the angle between its projection on the xy-plane and the x-axis. The vector $\mathbf{S}$ can be transformed to vector $\mathbf{S}'$ by the *rotation*

$$\mathbf{S}' = [\mathbf{M}]\mathbf{S} \tag{45}$$

where the matrix $\mathbf{M}$ is given by

$$\mathbf{M} = \begin{pmatrix} \cos\theta\cos\phi & \cos\theta\sin\phi & -\sin\theta \\ -\sin\phi & \cos\phi & 0 \\ \sin\theta\cos\phi & \sin\theta\sin\phi & \cos\theta \end{pmatrix}. \qquad (46)$$

The vector $\mathbf{S}'$ is along the z-axis and, thus, satisfies the equal uncertainty condition in the Eq. (42). As the θ and ϕ are varied, the spin vector $\mathbf{S}'$ traces out a sphere, know as the Bloch Sphere. That is, the tip of the vector always remains on the surface of the sphere as it should since the norm of the vector does not change with rotations. Hence, the atomic coherent states are also know as Bloch states and are represented by $|\theta, \phi\rangle$.

Let us now get back to the Eq. (42) giving the noise in an atomic coherent state. The atomic system being in this state, if we are able to squeeze out the noise in the x-component (say), that is

$$\langle S_x \rangle < \frac{\sqrt{N}}{2} \qquad (47)$$

at the expense of the other component such that the Eq. (41) is satisfied (or the vice-versa), then the system is said to be *SPIN SQUEEZED*. Thus, we realize from the above discussions that to find out if $\mathbf{S}$ is squeezed, we have to follow the following two steps:

1. Make the rotation as given in the Eq. (45).

2. Then, ask the question if $\mathbf{S}'$ satisfies the condition in the Eq. (47).

There are various nonlinear optical processes in which the atomic polarization P as a function of the electric field E of the incident field is given by

$$P = \chi_1 E + \chi_3 E^3 + \cdots \qquad (48)$$

can produces spin squeezed states. One has to look for a large χ_3. We have studied the Raman process (This process involves absorption of a photon from the incident laser called the pump and subsequent emission of a Stokes photon. This connects an upper intermediate level $|i\rangle$ to two low lying levels $|a\rangle$ and $|b\rangle$ of the interacting atom.) in which the the pump and the Stokes photon may have equal frequencies. Our investigation involved searching for regions where χ_3 was large and this could squeeze out a signification amount of noise in the spin system representing the atomic energy levels $|a\rangle$ and $|b\rangle$ [6]. A two-level system can also be in a spin squeezed state if it interacts with a *"tailored"* radiation field medium known in the language of quantum optics as squeezed vacuum. (There are quite a large number of papers on this topic and the reader may consult the Ref. [5] for further reading.) This interaction can be represented by the eigen value equation [7]

$$\Lambda|j,m,\xi\rangle = (S_x \cosh\xi + iS_y \sinh\xi)|j,m,\xi\rangle = m|j,m,\xi\rangle \qquad (49)$$

where the parameter ξ represents the surrounding with which the two-level system of N atoms is interacting. ($\xi = 0$ represents ordinary vacuum.) Also, $j = N/2$ with $-j \leq m \leq +j$. We notice that Λ is not Hermitian, that is, $\Lambda \neq \Lambda^\dagger$. But, its eigen values are real. This is due to the fact that Λ enjoys the property of pseudo-Hermicity [8] which means that $\Lambda^\dagger = \eta\Lambda\eta^{-1}$ and in addition $\eta = OO^\dagger$ with $O = O^\dagger = e^{-\xi J_z}$. We have been studying this problem and our initial results have the wave function for $m = 0$,

$$|j,\xi\rangle = \mathcal{N} \sum_{n=0}^{j} (-1)^n \frac{\sinh^n \xi}{n!} \sqrt{\frac{(j+n)!}{(j-n)!}} |j,n\rangle. \qquad (50)$$

This shows significant amount of spin squeezing for $\xi \neq 0$ [9]. In addition, there are a few papers which discuss spin squeezing using various atom-field(s) interactions. The spin squeezing has been experimentally observed [10].

VI. Conclusion

Thus we conclude that the noise can be reduced below than the minimum uncertainty noise in photonic as well as atomic systems.

In other words, the quantum mechanics can be tricked without violating the uncertainty principle, or for that matter, none of its rules or principles. We have noticed in the discussion that this was possible all due to the availability of highly stabilized lasers. Highly stabilized means that the radiation field from the laser is in a coherent state, or at least very close to a coherent state. Also, the lasers could produce large nonlinearities in its interactions with the material medium and thus producing squeezed states. We have seen in our study [6] that interaction of laser field with the atomic medium first squeezes the radiation field which, in turn, squeezes the two-level systems producing spin squeezing, a process know as self squeezing. Like this, this topic has many interesting results which is beyond the scope of this article. However, this paper gives an idea about the physics of squeezing and with the literature cited in the references the reader would be able to get deeper into the subject.

References

[1] E. Schrödinger, Naturwiss. **14**, 664 (1926).

[2] R. J. Galuber, Phys.Rev.Lett. **10**, 84 (1963); E. C. G. Sudarshan, Phys. Rev. Lett. **10**, 277 (1963).

[3] F. T. Arecchi, E. Courtens, R. Gilmore, and H. Thomas, Phys. Rev. A **6**, 2211 (1972).

[4] See, for example, L. A. Wu, M. Xiao, and H. J. Kimble, J. Opt. Soc. Am. B **4**, 1465 (1987). The first experimental observation was reported by R. E. Slusher, L. W. Hollberg, B. Yurke, J. C. Mertz, and J. F. Valley, Phys. Rev. Lett. **22**, 2409 (1985).

[5] *Squeezed states of the electromagnetic field*, Feature issue, J. Opt. Soc. Am B **4**, 1465 (1987).

[6] A. Dantan, M. Pinard, V. Josse, N. Nayak, and P. R. Berman, Phys. Rev. A **67**, 045801 (2003). Spin squeezing in a two-level system in a similar setup has been studied by L. Vernac, M. Pinard, and E. Giacobino, Phys. Rev. A **62**, 063812 (2000).

[7] G. S. Agarwal and R. R. Puri, Phys. Rev. A **49**, 4968 (1994); Also see D. J. Wineland, J. J. Bollinger. W. M. Itano, F. L. Moore, and D. J. Heinzen, Phys. Rev. A **46**, R6797 (1992).

[8] C. M. Bender and S. Boettcher, Phys. Rev. Lett. **80**, 5243 (1998); A. Mostafazadeh, J. Math. Phys. **43**, 205 and 2814 (2002).

[9] N. Nayak, R. N. Deb, and B. Dutta-Roy, In preparation.

[10] A. Kuzmich, K. Mølmer, and E. S. Polzik, Phys. rev. Lett. **79**, 4782 (1997); J. Hald, J. L. Sørensen, C. Schori, and E. S. Polzik, Phys. Rev. Lett. **83**, 1319 (1999); A. Kuzmich, L. Mandel, and N. P. Bigelow, Phys. Rev. Lett, **85**, 1594 (2000); C. Schori, B. Julsgaard, J. L. Sørensen, and E. S. Polzik, Phys. Rev. lett., **89**, 057903 (2002).

Physics of Solids, Nuclei and Particles
Editor: R. Sahu

The Nuclear Landscape

R.K. Choudhury

Institute of Physics, Bhubaneswar 751 005, India

ABSTRACT

All stable nuclei found in nature are arranged in a narrow band (β-stability line) in the nuclear chart close to $N=Z$ line. As one moves away from the beta stability line to the proton and neutron rich regions, many new features are being exhibited by the nuclear systems. Even in the stable peninsula, the nuclei exhibit shell and pairing effects, non-spherical shapes and deformations etc. in their ground state, which pose interesting challenges for a proper theoretical description of the nuclei. Recently, with the availability of radioactive nuclear beams from the accelerators, new frontiers of research have been opened up by producing nuclei with highly exotic combinations of N and Z across the nuclear chart. The possibility of synthesizing the doubly magic super heavy nuclei in the region of $Z=114{\sim}126$ and $N=170{\sim}184$ is now getting increasing realized. The present talk will cover certain specific areas of the vast field of nuclear physics research that is being carried out presently, in studying the behaviour of nuclei in different regions of mass and iso-spin space.

1.   <u>Introduction</u>

Nuclei are many body quantal systems consisting of neutrons and protons, which interact strongly with each other. The nature of nuclear interaction and the many body correlations between nucleons give rise to a variety of interesting features in nuclei in their ground states as well as excited states. The many different facets observed in nuclei such as the shell effects, magic numbers, deformed shapes, pairing and odd-even effects are unique to these systems. Although, there has been significant progress in understanding the nuclear forces and the behavior of nuclei in different regions of temperature, angular momentum, mass and iso-spin etc., one is still far away from a consistent and full description of nuclear systems. There are interesting phenomena which are manifested in the ground state structures of not only the stable nuclei, but also in nuclei far away from the beta stability line and in regions of high spins and excitation energies. The observations of shape co-existence, super-and hyper-deformations, neutron and proton halos, proton radioactivity and emergence of new magic numbers in neutron- and proton-rich nuclei, possible formation of stable super heavy nuclei etc. have provided rich ground for fundamental studies in nuclear physics over the last many decades. With the setting up of a number of large accelerator facilities with stable and radioactive ion beam species world wide, the expectations for accessing the nuclei in extreme neutron and proton-rich regions and in the super heavy mass region of $Z \geq 110$ are getting realized. In what follows, we will describe the interesting phenomena that are being manifested in nuclei as one spans different regions across the nuclear $N - Z$ landscape.

2.   <u>Peninsula of Stable Nuclei and Nuclei far away from Stability:</u>

In nature, nuclei have been synthesized by various astrophysical processes such as thermonuclear fusion and s- , r-, rp-processes. The relative abundances of nuclides as a function of mass number indicate that there are large enhancements for mass numbers corresponding to various even Z – even N configurations and for magic numbers in Z and N corresponding to 2, 8, 20, 28, 40, 50, 82 and 126. Moreover, the number of stable isotopes are relatively more for the magic numbers. The stable nuclides follow a narrow band in the nuclear chart of Z versus N. All these features imply the special nature of the strong interactions that bind the nucleons inside the nucleus. The characteristics of nuclear force are (i) short range ($\sim$ 1fm), (ii) charge symmetric, (iii) spin dependent and (iv) strong character ($\sim$ 100 times the electromagnetic force). The nuclear masse is decided by the total binding energy of the nucleons inside the nuclear volume. The semi-empirical mass formula that was formulated very early by Bethe and Weizsacker, has been continuously improved to be able to predict the nuclear masses and their stability with greater accuracy. The mass of a neutral atom with Z protons and N neutrons ($N = A - Z$), is written as (Wapstra, Handbuch der Physik).

$$M_a(A,Z)c^2 \quad = \quad Z(m_p + m_e)c^2 + Nm_nc^2 - B.\,E.$$

$$= \quad Z(m_p + m_e)c^2 + Nm_nc^2 - a_vA + a_sA^{2/3}$$

$$+ \; a_c(Z^2/A^{1/3}) + a_{asy}\{(A-2Z)^2\}/A + \delta/A^{1/2}$$

$$+ \; \Delta_{shell}$$

Where $a_v \quad = 15.835\ MeV$

$a_s \quad = 18.33\ MeV$

$a_c \quad = 0.714\ MeV$

$a_{aby} \quad = 23.20\ MeV$

$\delta \quad = 11.2\ MeV$ *for odd nuclei,*

$\quad\quad = 0\ MeV$ *for odd-even nuclei*

and $\quad = -11.2\ MeV$ *for even-even nuclei.*

Δ_{shell} *is a correction to account for the shell effects in nuclei.*

This formula provides a good description of the beta-stability valley, which can be written as

$$Z_{beta} = \frac{(4a_s + (m_n - m_p - m_e)c^2)A}{2(4a_s + a_cA^{2/3})}$$

The nuclear shell correction energies can be are obtained by the microscopic-macroscopic model, in which the difference in the energies of the realistic single particle levels and a uniform level scheme is taken to give the shell correction. In this way, the nuclear masses can be calculated for all the nuclei near the beta stability line with an accuracy of $1-2$ MeV. The mass formula, however, fails to account for the masses of nuclei lying very far away from the beta stability line.

It is also observed that most of the stable nuclei possess large non-zero quadrupole moments, implying significant deformations. The nuclear shell model with deformed harmonic oscillator potential is fairly successful in describing the single particle features of the nuclei and the ground state properties such as the magnetic and quadrupole moments and shell corrections. When one considers nuclei with either neutron-rich or proton-rich configuration, the shell model is not quite sufficient to describe the nuclear properties. For example, the various features of neutron **halo nuclei** (^{11}Li, ^{14}Be, ^{19}B, ^{23}C), vanishing of shell stability in nuclei with large proton and neutron excess, proton radioactivity in nuclei very far away from the beta-stability line etc. are difficult to describe in these models.

Study of $N=Z$ nuclei in the medium mass region ($A=60{\sim}140$) is another quite interesting problem from the point of understanding the charge symmetry effects, and the role of the asymmetry term in the mass formula. The level structures of the low lying excited states of many $N=Z$ nuclei ranging from $Z=32, 34, 36, 38, 40, 42, 44$ etc. have been investigated and it is observed that the collectivity in these nuclei is quite different as compared to the nuclei in the stable region for these Z numbers.

Another interesting aspect in the study of nuclei far away from the beta-stability line is the re-emergence of shell stability in the extreme neutron-rich nuclei in the medium mass region. Such effects are important from the point of r-process nucleo-synthesis for production of heavy elements. Some recent studies show that highly neutron-rich nuclei with ($Z=62, N=100$), ($Z=78, N=150$), ($Z=90$,

N=164) could have doubly shell closure configuration, providing extra stability to nuclei in these mass regions.

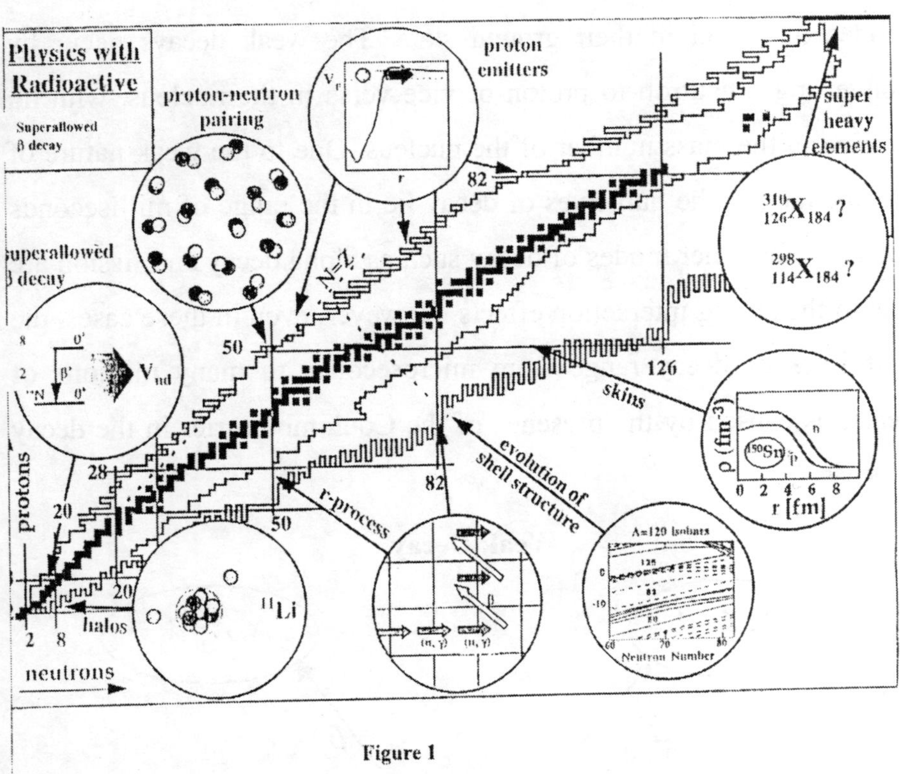

Figure 1

Presently, there is a lot of interest and effort to establish accelerator facilities to produce beams of radioactive nuclei away from the stability region. These accelerated radioactive nuclear beams will help in the study of various problems in nuclear physics and astrophysics areas as illustrated in Figure.1. Many countries including India are having programmes to set up major accelerator facilities in this regard.

3.   <u>Nuclear Decays :</u>

Fig. 2 gives a schematic representation of the various decay modes of nuclei in their ground state. The weak decays occur by converting a neutron to proton or vice versa in the nucleus, with no change in the mass number of the nucleus. Due to the weak nature of the interaction the half-lives of decay lie in the range of milliseconds to days. The other modes of decay such as alpha decay and fission are due to the strong interaction effects. However, even in these cases, the half lives of decay range from microseconds to many millions of years, as caused by the presence of the Coulomb barrier in the decay process.

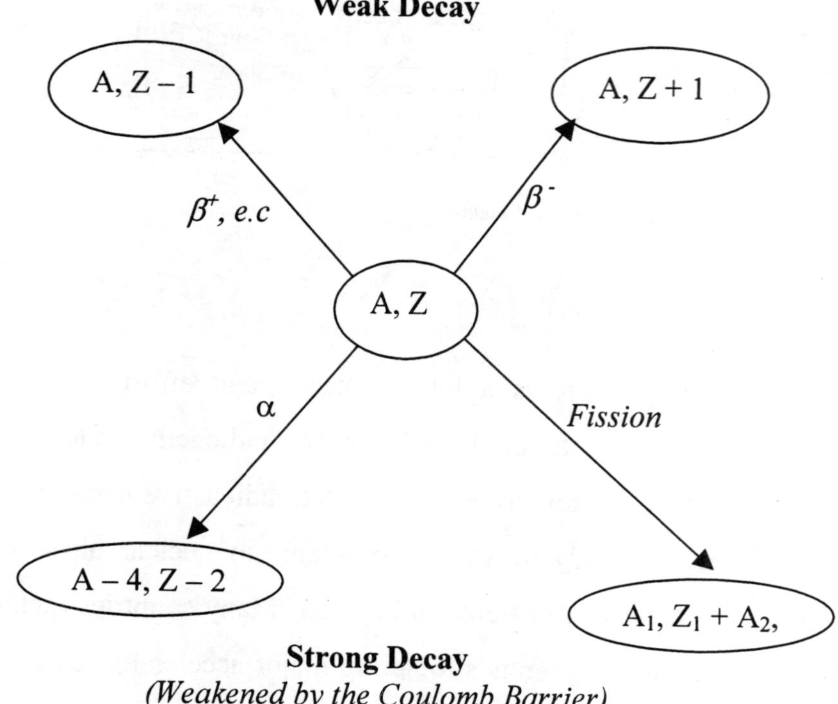

Strong Decay
(Weakened by the Coulomb Barrier)

Figure. 2.

The quantum mechanical tunnelling through this barrier gives rise to a strong dependence of the half-lives on the decay energy, as manifested by the Geiger-Nuttal law for the alpha decay process:

$$\log_{10}\lambda = C - \frac{D}{\sqrt{E_\alpha}}$$

where λ is the decay constant, C and D are nearly constant with respect to Z and N of the nucleus. The penetration probability (P) through the Coulomb barrier can be written by the WKB expression as :

$$P = \exp\left[\frac{2}{\hbar}\int_{R_{in}}^{R_{out}} \left\{\frac{2m_\mu m_f}{m_\mu + m_f}(V(r) - Q_\mu)\right\}^{1/2} dr\right]$$

where m_α and m_f are the mass of alpha particle and residual nucleus respectively. $V(r)$ is the potential energy as a function of radial distance of the alpha particle from the center of the residual nucleus. Q_α is the decay energy. R_{in} and R_{out} are the inner and outer radii corresponding to $V(r) = Q_\alpha$.

The decay constant , λ can be written as

$$\lambda_\alpha = N_{coll} \cdot P$$

where N_{coll} is the number of collision attempts made by alpha particle to penetrate the barrier.

A recent fitting of the alpha decay half-lives for a large number of nuclei has led to an improved version of the Geiger-Nuttal expression as :

$$\log_{10} T_\alpha = \frac{(aZ+b)}{\sqrt{Q_\alpha}} + (cZ+d) + h$$

where the values of the constants are obtained as :

a = 1.55261, b = 0.73247, c = -0.21669, d = -31.9949,

and   h = 0   for even − even   nuclei

         = 0.5 for odd A nuclei

         = 1.1 for odd − odd nuclei

The collision term N_{coll} is related to d as

$$N_{coll} = 10^{-d} \sim 10^{32}.$$

The above derivation, although qualitative, provides a conceptual description of the alpha decay process.

For heavy nuclei beyond $Z=90$, apart from alpha decay, spontaneous fission becomes another important decay mode, in which the nucleus splits into mainly two large fission fragments. The half-lives for such a process varies as shown in Fig.3. This can also be understood on the basis of penetration through a fission barrier, as caused due to change in the potential energy with deformation of the nucleus.

As the Z of the nucleus increases, the fission barrier height decreases due to the interplay of Coulomb and surface energy differences, as the nucleus undergoes deformation during the fission process. This explains the general decrease in the half-life with mass number of the fissioning nucleus. During the late 1960's, it was established that the fission barrier of a nucleus develops multiple maxima and minima due to the oscillatory nature of the shell correction energy with change in deformation (double humped fission barrier). This led to the understanding of the occurrence of spontaneous fission isomers, sub-barrier resonance in fission excitation functions, angular distributions

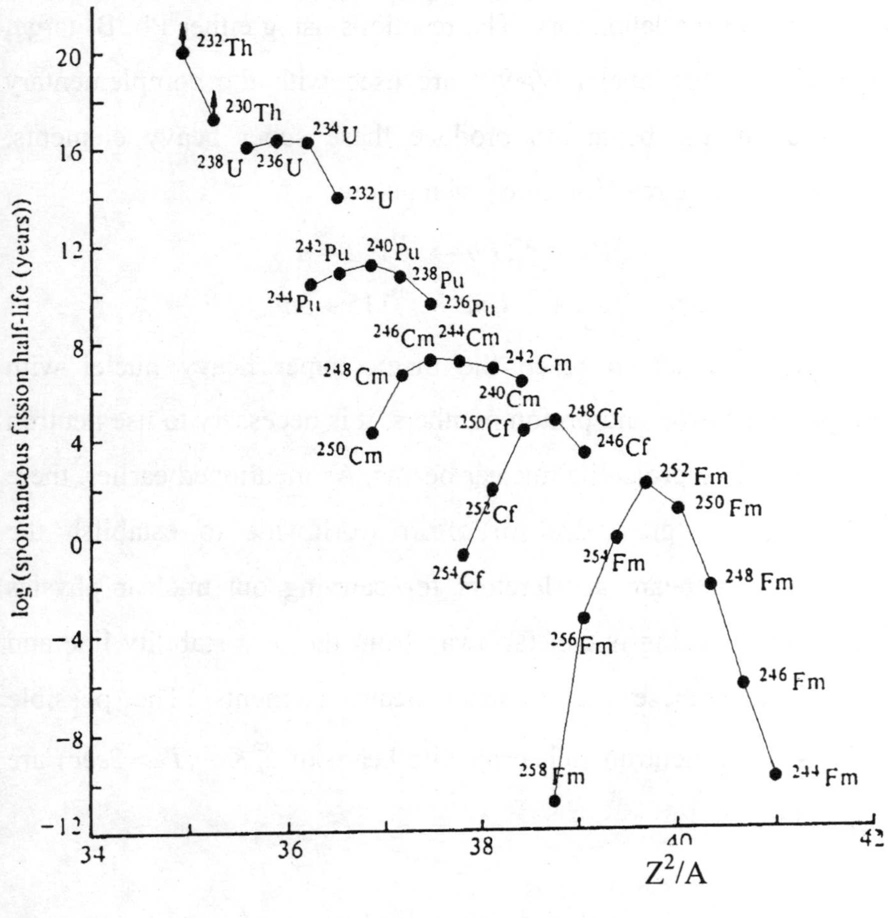

Figure 3

of fission fragments etc. Another important consequence of the shell effects is the persistence of the fission barrier for nuclei with large values of charge and mass beyond the actinide region. Thus, contrary to the liquid drop model predictions, there exists an island of nuclei beyond $Z=110$, which are stable against fission and alpha decay. The search for these super heavy elements has been pursued for last three decades, and there have been attempts also to synthesize them in the laboratory by means of heavy ion fusion reactions. So far the elements up to Z=113 and more recently, Z=115 have been synthesized in the laboratory. The reactions using either Pb, Bi target nuclei or actinide nuclei (Z>90) are used with the complementary projectile nuclear beams to produce these super heavy elements. Examples of these reactions are given below

$$_{30}^{70}Zn + \; _{82}^{208}Pb \rightarrow \; ^{277}112 + n$$

$$_{20}^{48}Ca + \; _{95}^{248}Cm \rightarrow \; ^{293}115 + 3n$$

However, in order to reach the magic super heavy nuclei with appropriate neutron and proton numbers, it is necessary to use neutron rich radio-active projectile nuclear beams. As mentioned earlier, there is at present a great deal of effort worldwide to establish the radioactive ion beam accelerators for carrying out nuclear physics research by reaching nuclei far away from the beta stability line and also to reach these magic super heavy elements. The possible reactions using neutron rich projectile beam of $_{36}^{93}Kr$ $(T_{1/2} \approx 2sec)$ are given below:

$$_{36}^{93}Kr + \; _{90}^{208}Pb \rightarrow \; ^{300}118 + n$$

$$_{36}^{93}Kr + \; _{90}^{232}Th \rightarrow \; ^{324}126 + n$$

It is to be seen whether such efforts will be successful. Nevertheless, the availability of a wide range of radioactive nuclear beams from accelerators will enable to access large regions of the nuclear landscape, which has remained unexplored so far.

4.   <u>Energy production in Stars and Nucleosynthesis.</u> :

Nuclear reaction processes form the basis of energy production in stars and synthesis of the heavy nuclei in stellar systems. The solar system is understood to have been formed *4.5 X 10⁹* years back. The proportion of the number of atoms of various elements in the solar system at the time of formation is as follows:

H	He	C	N	O	Ne	Mg	Si	S	Fe
2400	162	1.0	0.21	1.66	0.23	0.1	0.09	0.05	0.08

The total number of H atoms in the sun originally present is estimated to be about 9×10^{56}, out of which $\leq 10\%$ have been converted into heavier elements by the thermonuclear fusion processes during the lifetime of the sun. The various cycles for H-burning in the stars are given below :

i.   *proton – proton chain* :

Although an isolated proton can not undergo beta-decay to a neutron, the following reaction of fusing together two hydrogen nuclei becomes possible:

$$^1H + \,^1H \rightarrow \,^2H + e^+ + \nu + 0.42\text{MeV}$$

with a very small cross-section of about 10^{-27} barn at *5keV*, corresponding to a typical central stellar temperature of 10^7K. With a steady supply of deuterium due to the above reaction, the following reactions occur with differing rates, all leading to the ultimate effective conversion of ^{1}H to ^{4}He with release of energy.

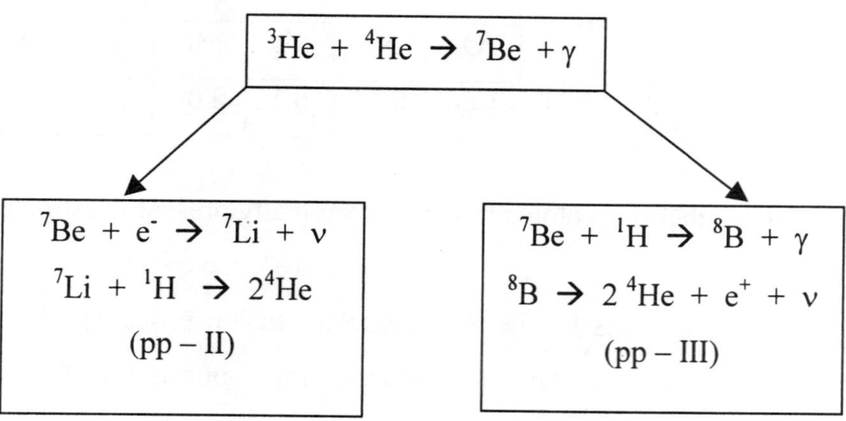

A total of about 25.0 *MeV* is released for conversion of 4^{1}H nuclei into one ^{4}He nucleus and two neutrinos are emitted during these processes. There has been, long outstanding problem for accounting the solar neutrino flux, since the observed neutrino flux detected in different laboratories can account for only 20% to 30% of that estimated from the solar models. The investigations in this direction have led to the discovery of solar neutrino oscillations, whereby an electron neutrino gets converted to a muon or tau neutrino, and this is possible only if neutrinos possess non-zero mass. Although a direct determination of neutrino mass is still a challenging task, these developments have opened up new frontiers of research in neutrino astronomy, nuclear and particle astrophysics etc.

ii.   *CNO Tri-Cycle* :

Another direction for the ^{1}H burning is through the so-called CNO-cycle, which can take place in the presence of small quantities of the stable isotopes of C, N and O along with the pre-dominant hydrogen. This route is expected to be important in freshly condensing later stage stars, where a mixture of heavy elements and hydrogen is expected to be present. The relative importance of the CNO-cycle in comparison with the pp-chains is greater with higher central temperature of the star.

Thermonuclear Fusion rates :

In the following, a simplistic general derivation of the fusion rates in the stars is given. Consider a volume of gas mixture of atoms of type a and A at a temperature T, with n_a and n_A nuclei per unit volume. For a given relative velocity, v between a and A, the reaction rate per unit volume is given by :

$$r = \sigma(v)\, v\, n_a n_A$$

for a velocity distribution, the average rate is

$$r = n_a n_A < \sigma v >$$

where

$$< \sigma v > = \int_0^\infty v\, \sigma(v)\, \Phi(v)\, dv$$

For Maxwellian distribution,

$$\Phi(v) = \sqrt{\frac{2}{\pi}} \left(\frac{\mu}{\kappa T} \right)^{3/2} v^2 \ \exp\left(-\frac{\mu v^2}{2\kappa T} \right)$$

so that

$$<\sigma v> \equiv \int_0^\infty \sigma(v)\,\Phi(v)\,v\,dv$$

$$= \left(\frac{8}{\pi\mu(\kappa T)^3}\right)^{1/2} \int_0^\infty \sigma(E)\,E\,\exp\left(-\frac{E}{\kappa T}\right)dx$$

where $E = \frac{1}{2}\mu v^2$, μ is the reduced mass. In case of highly sub-barrier reactions, due to pre-dominant quantum tunnelling effects, the energy dependent cross section $\sigma(E)$, can be expressed as :

$$\sigma(E) = \frac{S}{E}\,\exp\left(-\frac{b}{\sqrt{E}}\right)$$

S is called the "astrophysical S factor", and is nearly a constant quantity independent of E.

The average reaction rate is given by

$$r = n_a n_A \left(\frac{8}{\pi\mu(\kappa T)^3}\right)^{1/2} \int_0^\infty \exp\left(-\frac{E}{\kappa T} - \frac{b}{\sqrt{E}}\right)dE$$

The two terms of the integrand vary differently with energy as shown in Fig.4. The product shows a peak called "Gamow peak". at the energy :

$$E_0 = \left(\frac{b\,\kappa T}{2}\right)^{2/3}$$

which in case of pp reaction and for a central temperature of 10^7K ($\kappa T = 0.086 T_6\ keV$) turns out to be $\approx 5\ keV$.

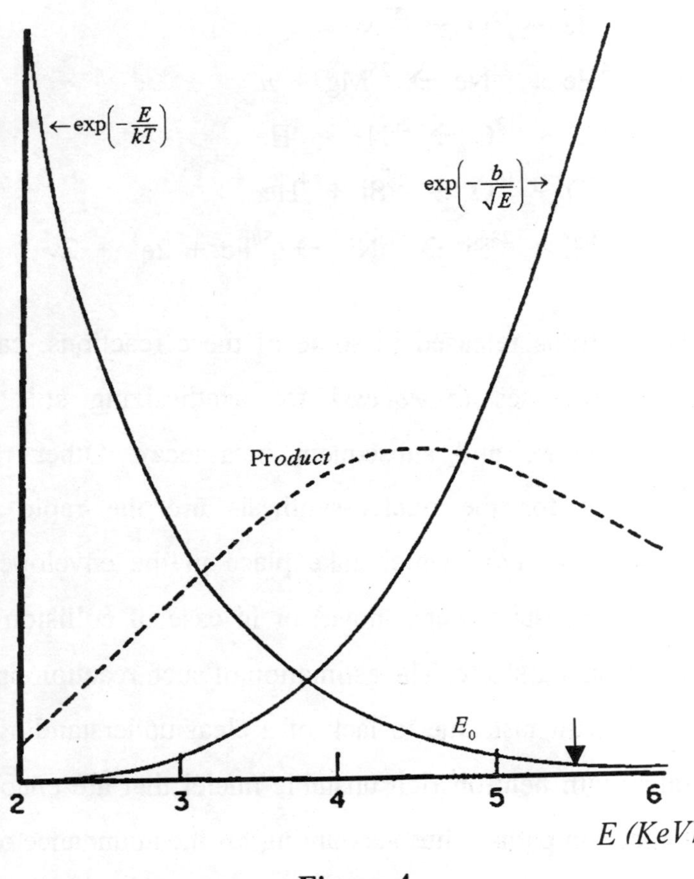

Figure 4

Thus, for a good understanding of the astrophysical processes, it is necessary to measure the nuclear cross sections at very low center of mass energies, which is by itself a very challenging task at present.

Nucleosynthesis Processes

Elements heavier than ^{4}He are synthesized through various routes, such as ^{4}He, ^{12}C, ^{16}O burning, as the temperature of the star interior rises to the range of 10^8K to 10^9K. The various nuclear reactions in this range are :

$$3\,^4\text{He} \rightarrow\ ^{12}\text{C} + \gamma$$

$$^4\text{He} + ^{14}\text{N} \rightarrow\ ^{18}\text{F} \rightarrow\ ^{18}\text{O} + e^+ + \nu$$

$$^{4}\text{He} + {}^{18}\text{O} \rightarrow {}^{22}\text{Ne}$$

$$^{4}\text{He} + {}^{22}\text{Ne} \rightarrow {}^{25}\text{Mg} + n$$

$$^{12}\text{C} + {}^{12}\text{C} \rightarrow {}^{20}\text{Ne} + {}^{4}\text{He}$$

$$^{16}\text{O} + {}^{16}\text{O} \rightarrow {}^{28}\text{Si} + {}^{4}\text{He}$$

$$^{28}\text{Si} + {}^{28}\text{Si} \rightarrow {}^{56}\text{Ni}^{*} \rightarrow {}^{56}\text{Fe} + 2e^{+} + 2\nu$$

The neutrons released in some of these reactions, cause the slow-capture processes (*s-process*) for synthesizing still heavier elements by capture and subsequent beta-decay. Other rare but important routes for the nucleo-synthesis are the rapid capture processes (*r-*, *rp-* *etc),* which take place in the envelope of the massive exploding stars (super-nova) or in case of collision of two massive "neutron stars" etc. The estimation of such reaction processes is a highly difficult task due to lack of a clear understanding of the reaction rates with neutron rich unstable nuclei that are encountered during the reaction paths. Thus accounting for the abundance of heavy elements, especially in the actinide region and in the super heavy region ($Z \geq 110$) is a very interesting and challenging problem of fundamental importance.

5.   <u>Summary</u> :

The aim of the present review has been to provide a glimpse of the various features of nuclear processes and the interesting phenomena that need to be studied for a proper understanding of the nuclei. The large variety of nuclear shapes and stability aspects of nuclei as we span the nuclear landscape, makes the study very challenging. In this review, we have selected only a few areas to highlight the progress made in this field. However, there are many directions in which nuclear physics research has advanced in studying the nuclear systems as function of excitation energy, angular momentum, iso-spin and deformation etc, which have not been covered in this review, and the reader is referred to many periodical articles appearing on these topics.

Physics of Solids, Nuclei and Particles
Editor: R. Sahu

Strange Matter in Astrophysics and Cosmology

Sibaji Raha

Department of Physics, Bose Institute, 93/1, A.P.C. Road, Kolkata 700 009, India

Over the past few decades, detailed and sensitive observations in the area of cosmology has enabled us to talk quite confidently of a *standard model* for the evolution of the universe, the so-called Big Bang Model. A detailed discussion of this model should, but unfortunately does not, form an integral part of any post-graduate curriculum in Physical Sciences. A full introduction to the topic in front of this audience would be very tempting and worthwhile but the time surely does not permit it. For our present purpose, then, let us simply summarize our present understanding, as derived from the recent WMAP (Wilkinson Microwave Anisotropy Probe) data [1].

WMAp data say that the Universe is 13.7 billion ($i.e. 13.7 \times 10^9$) years old ($\pm 1\%$).   The first stars ignited about 200 million years after the Big Bang. The content of the Universe is 4% Atoms (the stuff the usual world is made of), 23% Cold Dark Matter (matter responsible for clumping, structure formation and so on) and the remaining 73% is a mysterious smooth energy component, called the Dark Energy. (The dark energy is responsible for the recently observed accelerated expansion of the universe. More about this later on.) The expansion rate of the universe is denoted through the Hubble constant, $H_0 = 71$ km/sec/Mpc ($\pm 5\%$), the subscript 0 stands for the present day value.

One of the abiding mysteries in the Big Bang model is the nature of the Cold Dark Matter (CDM). Speculations as to the nature of dark matter are numerous, often bordering on the exotic, and searches for such exotic matter is a very active field of astroparticle physics. In recent years, there has been experimental evidence [2, 3] for at least one form of dark matter - the Massive Astrophysical Compact Halo Objects (MACHO) detected through gravitational microlensing effects proposed by Paczynski some years ago [4]. Based on about 13 - 17 Milky Way halo MACHOs detected in the direction of LMC - the Large Magellanic Cloud (we are not addressing the events found toward the galactic bulge), the MACHOs are expected to be in the mass range (0.15-0.95) $M_\odot$, with the most probable mass being in the vicinity of 0.5 $M_\odot$ [5], substantially higher than the fusion threshold of 0.08 $M_\odot$. To date, there is no clear picture as to what these objects are made of. I would like to present here a viewpoint that has emerged from the work of the Kolkata group over the past decade or so, that there is no need to go beyond the standard picture of particle interactions to understand the nature of CDM (or even Dark Energy (DE)).

Within the standard model of the evolution of the universe, the so-called Big Bang model, the succession of phase transitions plays a very important role. The phase transition occurring in the Grand Unified Theory (GUT) at an energy scale of 10^{15-16} GeV is believed to be responsible for the *inflationary* epoch in the early history of the universe. The electroweak phase transition at a few hundred GeV scale is expected to play a major role in baryogenesis. The next one in the sequence of such phase transitions is that from quarks to hadrons, the so-called QCD phase transition. Compared to the two phase transitions mentioned above, comparatively little attention has been paid to the QCD phase transition in the cosmological context, although intense efforts, both theoretical as well as experimental, have been devoted to the possibility of studying them in the laboratory, through energetic nuclear collisions. (It is a matter of some pride for us that the efforts of Indian scientists in this area have been seminal.) Given the enormous complexity of the system, there remains considerable ambiguity in the community about the order of the QCD phase transition. Indeed, it is still an open question whether the concept of thermodynamic equilibrium is valid in the short time scales of energetic nuclear collisions. In the cosmological scenario, however, the question of thermodynamic equilibrium does not pose a major problem; the time scale of Hubble expansion is certainly long enough, compared to the elementary interaction rates, to ensure thermal equilibrium among the species of interest during the epoch (a few microseconds after the Big Bang) where the QCD phase transition is expected to occur. In such circumstances, the cosmological QCD phase transition could indeed be of first order, especially as the size of the universe at that epoch is many orders of magnitude larger than the length scales associated with QCD interactions. Thus, the finite size effects would not camouflage a true first order phase transition. The cosmological implications of this, *presumably first order*, phase transition could indeed be far-reaching.

Among the various ramifications of a first order cosmic QCD phase transition, we shall, for the present purpose, confine our attention to only a few. A first order phase transition scenario involving bubble nucleation at a critical temperature $T_c \sim 100 - 200$ MeV should lead to the formation of quark nuggets (QN) [6], made of u, d and s quarks at a density $\geq$ nuclear density. If these primordial QN's existed till the present epoch, they could be possible candidates for the cold dark matter [6]. Such a possibility would be aesthetically rather pleasing, as it would not require any exotic physics nor would the success of the primordial (Big Bang) nucleosynthesis scenario be materially affected [7, 8, 9].

The central question in this context then is whether the primordial QN's can be stable on a cosmological time scale. Using the chromoelectric flux tube model, we have demonstrated [10] that the QN's will survive against baryon evaporation, if the baryon number of the quark matter inside the nugget is larger than 10^{42}. For reasons explained in [10], this estimate is rather conservative. Sumiyoshi and Kajino [11] have estimated that a QN with an initial baryon number $\sim 10^{39}$ would survive against baryon evaporation. It should be noted at this point that the horizon limit on the baryon number at that primordial epoch is around 10^{49}. Thus the current consensus in the community is that the larger primordial QN's within the baryon number window $10^{39-40} \leq N_B \leq 10^{49}$ are indeed cosmologically stable. (Also noteworthy in this context is the observation that these numbers are sufficiently above the limit ($N_B \sim 10^{26}$) which was obtained by Madsen [12] some time ago, below which the QN's could interfere with the outcome of Big Bang nucleosynthesis.) It is therefore most relevant to ask what fraction of the dark matter could be accounted for by the surviving QN's [13].

It is well known that in a first order phase transition, the quark and the hadron phases co-exist. This configuration would be referred to in the following as the mixed phase. In the quark phase, the universe consists of leptons, photons as well as the quantum chromodynamic (QCD) degrees of freedom (massless quarks, antiquarks and gluons) and is described by, say, the MIT bag equation of state with an effective degeneracy $g_Q \approx 51.25$. Note that the baryon number in this phase is carried entirely by the quarks. The hadronic phase contains baryons, mesons, photons and leptons and is described by an equation of state corresponding to massless particles with an effective degeneracy $g_H = 17.25$. The evolution of the universe in the mixed phase at the critical temperature T_c of the phase transition is governed by the Einstein equation in the Robertson-Walker space time, as described below.

$$\left(\frac{\dot{R}}{R}\right)^2 = \frac{8\pi\epsilon}{3m_{pl}^2} \tag{1}$$

$$\frac{d(\epsilon R^3)}{dt} + P\frac{dR^3}{dt} = 0 \tag{2}$$

where ϵ is the energy density, P the pressure and m_{pl} the Planck mass. Combining eqs. (1) and (2) with the equation of state, we determine the scale factor $R(t)$ and the volume fraction of the quark matter $f(t)$ in the mixed phase as

$$\frac{R(t)}{R(t_i)} = \frac{\left[cos\left(arctan\sqrt{3r} - \sqrt{\frac{3}{r-1}}(t-t_i)/t_c\right)\right]^{2/3}}{\left[cos\left(arctan\sqrt{3r}\right)\right]^{2/3}} \tag{3}$$

and

$$f(t) = \frac{1}{3(r-1)}\left[tan\{arctan\sqrt{3r} - \sqrt{\frac{3}{r-1}}\frac{t-t_i}{t_c}\}\right]^2 - \frac{1}{r-1} \tag{4}$$

where $r \equiv g_q/g_h$, $t_c = \sqrt{3m_{pl}^2/8\pi B}$ is the characteristic time scale for the QCD phase transition in the early universe and t_i is the time when phase transition starts. B is obviously the vacuum energy density of the bag, called the Bag constant.

In the mixed phase, the temperature of the universe remains constant at T_c, the cooling due to expansion being compensated by the liberation of the latent heat. In the usual picture of bubble nucleation in first order phase transitions, hadronic matter starts appearing in the quark matter as individual bubbles. With the progress of time, more and more hadronic bubbles form, coalesce and eventually percolate to form a network of hadronic matter which traps the quark phase into finite domains [14]. The time when the percolation takes place is called the percolation time t_p, determined by a critical volume fraction f_c, $(f_c \equiv f(t_p))$ of the quark phase.

Detailed numerical studies on percolating systems yield the result that for bubbles with the same radial size, f_c is ~ 0.3 [15]. We would also use the same value of f_c here. For the sake of simplicity, we would also assume that the trapped quark domains are all of the same size. (In general, this is not true; there ought to be a distribution of domain sizes. We have recently studied the size distribution

                    Physics of solids, nuclei and particles

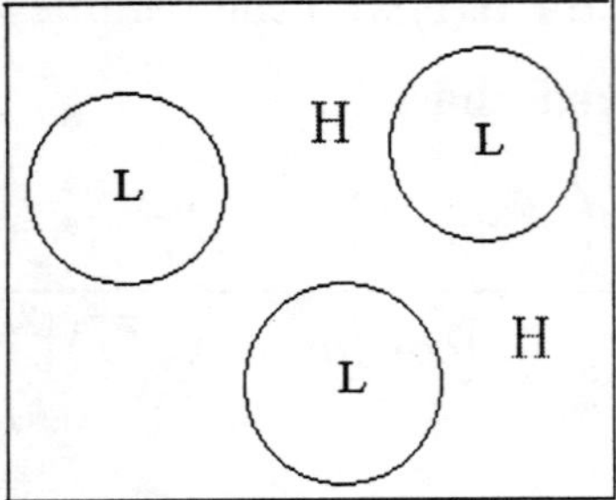

Isolated expanding bubbles of low temp
in high temp. phase

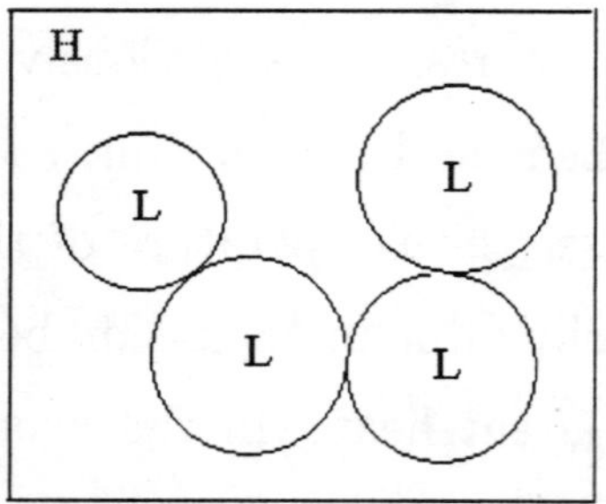

Expanding bubbles meet

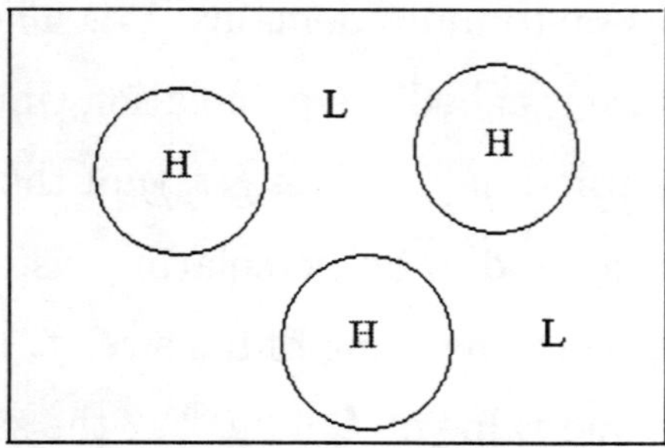

Isolated shrinking bubbles of High temp phase

Figure 1: Schematic diagram of a first order cosmic QCD transition

of such domains and found [16] that for all reasonable scenarios, the distribution is rather sharply peaked. Thus, the assumption is quite reasonable.)

From eq.(4), we get $t_p - t_i = 0.08t_c$. For $B^{1/4} = 245$ MeV, we obtain $t_c = 46\mu$sec, $t_p = 27\mu$sec and $t_i = 23\mu$sec.

In an ideal first order phase transition, the fraction of the high temperature phase decreases from the critical value f_c, as these domains shrink. For the QCD phase transition, however, these domains should become QN's [6] and as such, we may safely assume that the lifetime of the mixed phase t_f (*i.e.*, the time when the cooling due to expansion starts to dominate again and the temperature of the universe starts falling), is $\sim t_p$.

The probability of finding a domain of trapped quark matter of co-ordinate radius χ at time t_p is given by [17],

$$P(z, x_p) = \exp\left[-\frac{4\pi}{3}v^3 t_c^4 \int_{x_i}^{x_p} dx I(x) \left(zr(x) + y(x_p, x)\right)^3\right] \quad (5)$$

where $z = \chi R(t_i)/vt_c$, $x = t/t_c$, $r(x) = R(x)/R(x_i)$ and $I(x)$ is the rate of nucleation per unit volume. v is the radial growth velocity of the nucleating bubbles, which is left as a parameter. $y(x_p, x)$ is given by the following equation [17]

$$y(x, x\prime) = \int_{x\prime}^{x} r(x\prime)/r(x\prime\prime)dx\prime\prime \quad (6)$$

Various authors [14, 18, 19] have proposed different nucleation rates for the cosmic QCD phase transition. Let us start with the prescription of ref. [14], where the nucleation rate is given by the following expression

$$I(t) = r_T\delta(t - t_i) \quad (7)$$

where the prefactor of the thermal nucleation rate r_T is determined from the requirement $P(0, t_p) = f_c = 0.3$, yielding

$$r_T = \frac{3r_c}{4\pi v^3 t_c^4} \frac{1}{y(x_p, x_i)} \tag{8}$$

with $r_c \sim 1.2$. This leads to

$$P(z, x_p) = \exp\left[-r_c \left(\frac{zf(x_i)}{y(x_p, x_i)} + 1\right)^3\right] \tag{9}$$

The radius of the trapped quark domain may be taken to be the length scale where the probability $P(z, x_p)$ falls to f_c/e. This implies

$$\frac{r_{QN}}{vt_c} \approx 0.019 \tag{10}$$

The number density of the QN's (n_{QN}) can be obtained by using the relation $n_{QN} V_{QN} = f_c$ as

$$n_{QN} \approx \frac{418}{(vt_p)^3} \tag{11}$$

In an idealized situation where the entire CDM universe is due to the QN's, we should have,

$$N_B^H(t_p) = N_B^{QN} n_{QN} V^H(t_p) \tag{12}$$

where $N_B^H(t_p)$ is the total number of baryons required provide ($\Omega_{CDM} = 0.23$) at t_p, N_B^{QN} is the total number of baryons contained in a single quark nugget and $V_H(t_p) = 4\pi(ct_p)^3/3$ is the horizon volume.

Demanding that $v/c \leq 1/\sqrt{3}$ (*i.e., the bubble growth velocity is subluminal*), we get

$$N_B^{QN} \leq 10^{-4} N_B^H(t_p) \tag{13}$$

Since the usual baryons constitute only $\sim 4\%$ of the closure density, a total baryon number of 10^{50} in the form of QN's within the horizon at a temperature of ~ 100 MeV would accommodate all the CDM. This would require N_B^{QN} to be $\leq 10^{46}$, which is within the survivability limit of QN's mentioned earlier.

For the sake of completeness, we would like to mention here that QN's with baryon number lower than the survivability window would evaporate rather quickly, leaving a large baryon inhomogeneity. This would still not pose much problem for the cosmological scenario, as this overdensity would dissipate, primarily due to neutrino inflation upto temperatures ~ 1 MeV and then by baryon diffusion, which becomes the dominant mechanism at lower temperatures. Even if the initial overdensity due to the evaporating QN's could be as high as 10^{10-12}, it is found to go down by several orders of magnitude by the time nucleosynthesis starts [16].

It is therefore most relevant to investigate the fate of these SQNs. Recalling that the number distribution of the SQNs is sharply peaked [16], we shall assume that all the SQNs have the same baryon number.

These SQN's, if formed during the cosmic QCD phase transition at T$\sim$100 MeV, behave like very unusual cold dark matter objects. Their enormous mass ($\sim 10^{44}$ GeV) and large size ($R_N \sim 1$m) force them to be almost static objects in the coordinate space. Even if they continue to be in kinetic equilibrium due to the radiation pressure (photons and neutrinos) acting on them, their velocity would be extremely non relativistic. Also their mutual separation would be considerably larger than their radii; for example, at T $\sim$ 100MeV, the mutual separation between the SQN's is (of size $\sim 10^{44}$ baryons) estimated to be around ~ 300m. It is then obvious that the SQN's do not lend themselves to be treated in a hydrodynamical frame-work; they behave rather like discrete bodies in the background of the radiation fluid. They thus experience the radiation pressure, quite substantial because of their large surface area as well as the gravitational potential due to the other SQN's.

In such circumstances, it would be rather tempting to use the powerful tool of the Virial theorem [20, 21]. Even a naive application of the Virial theorem implies that the internal energy $\mathbf{U}$ of all the SQN's contained within the event horizon $(\sim 2t)$ at any instant of time is substantially less than $\frac{|E|}{2}$, E being the total gravitational energy, so that these SQN's would tend to undergo gravitational collapse. Such, of course, would not be the case for any other massive particles like baryons; their kinetic energy would continue to be very large till very low temperatures. More seriously, the Virial theorem can be applied only to systems whose motion is sustained. In the presence of dissipation killing all kinetic energy, the Virial theorem becomes inapplicable. SQN's have a very unusual property in that they tend to absorb baryonic matter without limit, becoming more and more bound in the process. They would therefore clump together whenever two such SQN's collide, damping their motion even further.

We therefore have to estimate to what extent the radiation pressure can prevent the SQN's from gravitating towards one another. It should be mentioned at this juncture that for the system of discrete SQN's suspended in the radiation fluid, a detailed numerical simulation would be essential before any definite conclusion about their temporal evolution can be arrived at. This is a quite involved problem, especially since the number of SQN's within the event horizon, as also their mutual separation, keeps increasing with time. Our purpose in the present work is to examine whether such an effort would indeed be justified.

Let us now consider the possibility of two nuggets coalescing together under gravity, overcoming the radiation pressure. The mean separation of these nuggets and hence their gravitational interactions are determined by the temperature of the universe. If the universe is assumed to be closed with SQNs, the total baryon number contained in them within the horizon at the QCD transition temperature (~ 100 MeV) would be 10^{51}. For SQNs of baryon number b_N each, the number of SQNs within the horizon at that time would be just $(10^{51}/b_N)$. Then at any later time, the number of SQNs within the horizon (N_N) and their density (n_N) as a function of temperature would be given by :

$$N_N(T) \;=\; \frac{10^{51}}{b_N}\left(\frac{100\text{MeV}}{\text{T}}\right)^3 \tag{14}$$

$$n_N(T) \;=\; \frac{N_N}{V_H} = \frac{3N_N}{4\pi(2t)^3} \tag{15}$$

where the time t and the temperature T are related in the radiation dominated era by the relation :

$$t = 0.3 g_*^{-1/2}\frac{m_{pl}}{T^2} \tag{16}$$

with g_* being ~ 17.25 after the QCD transition [13].

From the above, it is obvious that the density of SQNs decreases as $t^{-3/2}$ so that their mutual separation increases as $t^{1/2}$. Therefore, the force of their mutual gravitational pull will decrease as t^{-1}. On the other hand, the force due to the radiation pressure (photons and neutrinos) resisting motion under gravity would be proportional to the radiation energy density, which decreases as T^4 or t^{-2}. It is thus reasonable to expect that at some time, not too distant, the gravitational pull would win over the radiation pressure, causing the SQNs to coalesce under their mutual gravitational pull.

The force due to the radiation pressure on the nuggets may be roughly estimated as follows. We consider two objects of the size of the nugget's approaching each other due to gravitational interaction, overcoming the repulsion due to the radiation pressure. The usual isotropic radiation pressure being $f\rho c^2$ where ρ is the radiation density, and f is the degeneracy factor, the nuggets will have to overcome an additional pressure gradient due to their mutual motion, which is $f\rho c^2(\gamma - \frac{1}{\gamma})$, where $\beta = \frac{v_{\text{fall}}}{c}$) is the velocity of approach, essentially due to the fact that the volumes are contracted and dilated in the direction of motion and opposite to the direction of motion respectively. This simplifies to $\frac{5}{6}\rho_{\text{rad}}cv_{\text{fall}}\beta\gamma$ with the insertion of the appropriate degeneracy factor. Hence the expression for the force can be written as,

$$F_{\text{rad}} = \frac{5}{6}\rho_{\text{rad}}cv_{\text{fall}}(4\pi r^2)\beta\gamma \tag{17}$$

where ρ_{rad} is the radiation density (including both photons and neutrinos), v_{fall} or βc is the velocity of SQNs determined by mutual gravitational field and γ is $1/\sqrt{1 - \beta^2}$. The quantities F_{rad}, β and γ all depend on the temperature of the epoch under consideration. (It is worth mentioning at this point that the t dependence of F_{rad} is actually $t^{-5/2}$, sharper than the t^{-2} estimated above, because of the v_{fall}, which goes as $t^{-1/4}$.) The ratio of these two forces is plotted against temperature for two SQNs with initial baryon number 10^{42}

each. It is obvious from the figure that ratio $F_{\text{grav}}/F_{\text{rad}}$ is very small initially. As a result, the nuggets will remain separated due to the radiation pressure. For temperatures lower than a critical value T_{cl}, the gravitational force starts dominating, facilitating the coalescence of the SQNs under mutual gravity.

Table 1: Critical temperatures (T_{cl}) of SQNs of different initial sizes b_N, the total number N_N of SQNs that coalesce together and their total final mass in solar mass units.

b_N	T_{cl} (MeV)	N_N	$M/M\odot$
10^{42}	1.6	2.44×10^{14}	0.24
10^{44}	4.45	1.13×10^{11}	0.01
10^{46}	20.6	1.1×10^{7}	0.0001

Let us now estimate the mass of the clumped SQNs, assuming that all of them within the horizon at the critical temperature will coalesce together. This is in fact a conservative estimate, since the SQNs, although starting to move towards one another at T_{cl}, will take a finite time to actually coalesce, during which interval more SQNs will arrive within the horizon. In table 1, we show the values of T_{cl} for SQNs of different initial baryon numbers along with the final masses of the clumped SQNs under the conservative assumption mentioned above.

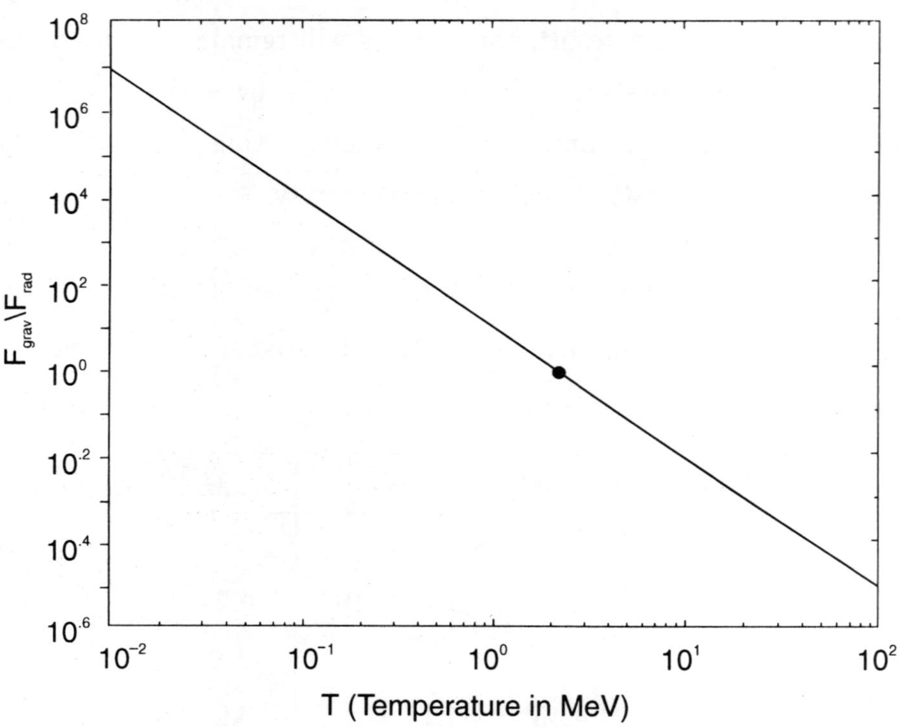

Figure 2: Variation of the ratio $F_{\text{grav}}/F_{\text{rad}}$ with temperature. The dot represents the point where the ratio assumes the value 1.

It is obvious that there can be no further clumping of these already clumped SQNs; the density of such objects would be too small within the horizon for further clumping. Thus these objects would survive till today and perhaps manifest themselves as MACHOs. It is to be reiterated that the masses of the clumped SQN's given in table 1 are the lower limits and the final masses of these MACHO candidates will be larger. A more detailed estimate of the masses will require a detailed simulation, but very preliminary estimates indicate that they could be 2-3 times bigger than the values quoted in table 1.

The total number of such clumped SQNs within the horizon today would be $\sim 10^{23-24}$. We can also mention at this juncture that if the MACHOs are indeed made up of quark matter, then they cannot grow to arbitrarily large sizes. Within the (phenomenological) Bag model picture [22] of QCD confinement, where a constant vacuum energy density (called the Bag constant) in a cavity containing the quarks serves to keep them confined within the cavity, we have earlier investigated [23] the upper limit on the mass of astrophysical compact quark matter objects. It was found that for a canonical Bag constant **B** of (145 MeV) ^{4}, this limit comes out to be 1.4 $M_\odot$. The collapsed SQNs are safely below this limit.

As a consistency check, we can perform a theoretical estimate of the abundance of such MACHO's which is conventionally given by the optical depth. The optical depth is estimated as [24]:

$$\tau = \frac{4\pi G}{c^2} D_s^2 \int \rho(x) x(1-x) dx \tag{18}$$

where the symbols have their meanings as given in the above reference. In particular ρ is the mass-density of the Machos, which is of the form $\rho = \rho_0 \frac{1}{r^2}$ in the naive spherical halo model. Assuming an $\Omega_B = 0.01$ for the usual *visible* baryons, the total mass of the Milky Way ($\sim 1.6 \times 10^{11} M_\odot$) corresponds to a factor of $\sim 2 \times 10^{-9}$ of all the baryons within the present horizon. Scaling the number of clumped SQNs within the horizon by the same factor yields a total number of $\sim 2 - 3 \times 10^{13}$ (for $n_N = 10^{42}$; for $n_N = 10^{44}$, this would become $\sim 3 - 5 \times 10^{14}$) in the Milky Way halo. For a naive inverse square spherical model comprising such objects upto the LMC, we obtain an optical depth of $\sim 2 - 7 \times 10^{-7}$. The uncertainty in this value is mainly governed by the value of Ω_B, and to a lesser extent by the specific halo model. This value compares reasonably well with the observed value and may be taken as a measure of reliability in the proposed model.

The issue of Dark Energy can also, intriguingly, solved within the same scenario. We have recently shown that by requiring an overall colour singlet state for the entire universe, the dark energy can emerge naturally from the cosmic QCD phase transition, from the quantum property of entanglement in the colour space. The shortage of time prevents me from discussing this issue at great depth; the interested reader may look at our recent preprint in the astro-ph archives.

The scenario proposed here would get substantial experimental support, albeit somewhat indirect, if a sizable flux of stable strangelets (heavy nuclear objects with very abnormal charge-to-mass ratios) could be detected in the cosmic rays. (These strangelets could arise from collision of strange stars or even strange nuggets.) To this end, it is necessary to understand the propagation of strangelets through the earth's atmosphere, as these would have to traverse the whole atmospheric depth before arriving at the detectors.

In heavy ion collisions, it is thought that strangelets with atomic number A upto 20 - 30 may be formed [25], the stability of which depend rather sensitively on the parameter values (like the Bag constant) and an underlying shell-like structure. For larger strangelets ($A > 40$), the stability appears to be more robust [26]. We confine our attention in this work to such larger strangelets, which may not be readily formed in heavy ion collisions in the laboratory but could be of cosmic origin. A discerning property of such strangelets would be an unusual charge to mass ratio ($Z/A \ll 1$) [26].

The obvious place to look for such strangelets would be in the cosmic ray flux. There have been intermittent reports in the literature [27, 28, 29, 30, 31] about the detection of exotic cosmic ray events, with unusually low charge to mass ratios. Some of these events are tabulated in Table 2.

Table 2: Mass and charge obtained from cosmic ray experiments

Event	*Mass*	*Charge*
Counter experiment [27]	$A \sim 350\text{-}450$	14
Exotic Track [28]	$A \sim 460$	20
Price's Event [29]	$A > 1000$	46
Balloon Experiments [30, 31]	$A \sim 370$	14

Although it appears natural to identify these events with strangelets, no consensus has yet emerged, primarily because of the ambiguities associated with the mechanism of propagation of strangelets through the terrestrial atmosphere. For example, if a strangelet arriving at the top of the atmosphere has a baryon number $A \sim 1000$, there would be a serious problem with its penetrability through the atmosphere, as the *exotic* events are observed at quite low altitudes. One could assume that their geometric cross sections are very small. Alternately, one could conjecture, *a la* Wilk *et al* [32, 33, 34] and others [35], that although the initial mass of the strangelet is very large, it decreases rapidly due to collisions with air molecules, until the mass reaches a critical value m_{crit} below which the strangelet simply evaporates into neutrons.

The difficulties associated with this kind of interpretation are twofold. Firstly, one has to take account of the fact that, unlike ordinary nuclear fragments which tend to break up in collisions, strangelets can become more strongly bound if they absorb matter [6]. Secondly, since a strangelet has a net electric charge, it experiences an ever increasing geomagnetic field, which considerably lengthens its path before reaching a certain altitude. This implies many more interactions with the nuclei of the atmospheric atoms, as a result of which the strangelet would "evaporate" much before the desired depth is reached.

These difficulties can be naturally overcome in a different scenario, proposed recently by the present authors [36], in which the stability of the strangelet plays a very important role. In this model, an initially small strangelet, during its travel through the Earth's atmosphere, picks up mass, rather than lose it, from the atmospheric atoms. Such a situation may prevail unless the propagation velocity of the strangelet through the atmosphere is so high that in a collision with the atmospheric nucleons, the excitation energy would exceed the binding energy. We have estimated that for our case, where the initial A is larger than 40, this upper limit on the velocity comes out to be above 0.7c. (We disregard the possibility of fission-like fragmentation of the strangelets.) The equation governing the rate of change of mass with respect to distance traveled is given by,

$$\frac{dm_s}{dh} = \frac{f \times m_n}{\lambda} \tag{19}$$

and the equation of motion reads

$$\frac{d\vec{v}}{dt} = -\vec{g} + \frac{q}{m_s}(\vec{v} \times \vec{B}) - \frac{\vec{v}}{m_s}\frac{dm_s}{dt} \tag{20}$$

In the above equation, m_s and $\vec{v}$ represent the instantaneous mass and the velocity of the strangelet, q is the charge and λ is the mean free path of the strangelet in the atmosphere. The factor f determines the fraction of neutrons that are actually absorbed out of the incident neutrons (m_n). In this case, λ is both a function of h (which determines the density of air molecules) and m_s (which is related to the interaction cross section). The initial velocity has to be bigger than a threshold value, so that a strangelet of a given initial mass and charge can arrive at an altitude $\sim 25\,km$ from the sea level, surmounting the geomagnetic barrier.The upper limit of 25 km is chosen primarily to economize on the computation time and is *a fortiori* justified since the density of the atmosphere above this height is almost negligible for our purpose. The variation of atmospheric density with height has been described by a parametric fit to the data given in the standard reference of Kaye and Laby [37].

According to the above analysis, a strangelet with an initial mass of $\sim 64\ amu$ and charge ~ 2 evolves to a mass $\sim 340\ amu$ or so, by the end of its journey, an altitude $\sim 3.6\ km$ above the sea level (typically the height of a northeast Himalayan peak in India, like *Sandakphu* , at a geomagnetic latitude $\sim 30^0\ N$). This mass is quite close to the few available data (see Table 2) and seems to support the interpretation that exotic cosmic ray events with very small Z/A ratios could result from SQM droplets. However, it was assumed in [36] that only neutrons are absorbed preferentially over the protons from the nuclei of the atmospheric atoms (*i.e.* charge of the strangelet remains constant), the protons being coulomb repelled. It should nonetheless be realized that in the earlier phase of the journey, when the relative velocity between the strangelet and the air molecule is large, some protons will indeed be absorbed, albeit with a lower cross section than that for neutron capture. As the strangelet builds up in mass as well as in charge, the coulomb barrier at the surface of the strangelet gets steeper and the relative velocity also gets further reduced. This will slow down the charge transfer process and ultimately inhibit it. Also, one cannot avoid the issue of loss of energy of the strangelet through ionization of the surrounding media. As we shall see, the ionization losses, which become quite significant at comparatively low altitudes, actually provides a lower limit to the height at which the strangelets can be detected successfully.

The equation of motion (20) can be generalized to a relativistic form in a straightforward manner:

$$\gamma m_s \frac{d\vec{v}}{dt} = -m_s \vec{g} + q(\vec{v} \times \vec{B}) - \gamma \vec{v} \left(\frac{dm_{s_n}}{dt} + \frac{dm_{s_p}}{dt} \right) - m_s \vec{v} \frac{d\gamma}{dt} - \frac{f(v)}{\sqrt{3}} \vec{v}$$

$$(21)$$

where γ is the Lorentz factor. The third term takes care of the deceleration of the strangelet due to the absorption of neutron as well as protons, where the proton absorption term is related to the neutron absorption term as

$$\frac{dm_{s_p}}{dt} = \frac{\sigma_p}{\sigma_n} \frac{dm_{s_n}}{dt} \equiv f_{pn} \frac{dm_{s_n}}{dt}$$

$$(22)$$

where σ_p and σ_n are the cross sections for neutron and proton absorption, respectively. Treating, classically, the proton of energy E as a free charged particle of unit charge in the repulsive coulomb field of the strangelet, we can easily estimate the minimum separation r_{min} along the trajectory to be given by

$$\frac{(mv_o b)^2}{2mr_{min}} + U(r_{min}) = E$$

where $U(r)$ represents the potential energy of the proton due to the coulomb field of the strangelet; v_o is the relative speed with which the N_2 nuclei (and hence, its constituent protons) approach the strangelet and b is the impact parameter. Assuming that charge transfer can take place when $r_{min} \leq R_s$ (the radius of the strangelet), the corresponding value of $b(\equiv b_c)$ is $b_c^2 = R_s^2(1 - U(R_s)/E)$, so that the proton capture cross section (σ_p) by a strangelet of atomic number Z_s is $\sigma_p = \pi b_c^2 = \pi R_s^2 \left[1 - \frac{Z_s e^2}{4\pi\epsilon_0 R_S} / E \right]$.

In contrast, the scattering cross section for neutrons (σ_n) is just $\pi(r_n + R_S)^2$ and hence,

$$f_{pn} = \frac{R_s^2}{(r_n + R_s)^2}\left(1 - \frac{1}{E}\frac{Z_s e^2}{4\pi\epsilon_0 R_s}\right) \tag{23}$$

Finally, the last term of equation (21) accounts for the ionization loss. The expression for $f(v)$ is given by [38]

$$f(v) = -\frac{dE}{dx} = \frac{Z_s^2 e^4 n Z_{med}}{4\pi\epsilon_0^2 m_e v^2}ln\left(\frac{b_{max}}{b_{min}}\right) \tag{24}$$

Here, n represents the number density of the atmospheric atoms at a particular altitude, Z_{med} is the number of electrons per atom of N_2 which can be ionized, m_e is the mass of the electron and b_{max} and b_{min} are the maximum and minimum values of the impact parameter. At large velocities, expression (24) reduces to, with I denoting the average ionizing energy,

$$f(v) = \frac{Z_s^2 e^4 n Z_{med}}{4\pi\epsilon_0^2 m_e v^2}ln\left(\gamma^2\frac{2m_s v^2}{I} - \beta^2\right) \tag{25}$$

However, when the velocity of the strangelet falls below a critical value $v \leq 2Z_s v_0$ ($v_0 = 2.2 \times 10^6 m/s$ is the speed of the electron in the first Bohr orbit), electron capture becomes significant which can be accounted for by the replacement $Z_s \to Z_s^{\frac{1}{3}}\frac{v}{v_o}$ [38, 39].

Equation (21) is solved by the 4th order Runge-Kutta method with different sets of initial mass, charge and β. It may be mentioned at this point that the first term in eqn (21) is not important in magnitude, as is to be expected. We nonetheless include it for numerical stability. This serves to define the downward vertical direction in the vector algorithm, especially for very small initial velocities.

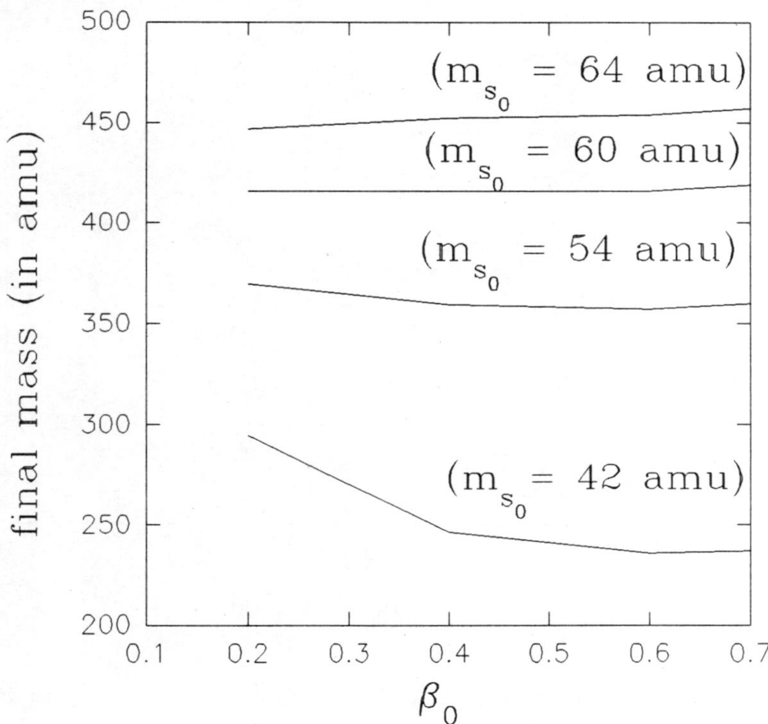

Figure 3: Variation of final masses with initial β for different initial masses

In figure 3, we have plotted final masses (for initial masses 42, 54, 60 and 64 *amu*) with initial β for a fixed initial charge 2. This graph shows that the final value of the mass decreases at first with increasing values of the initial β and then begins to increase again after a critical value of β is reached. This feature, although not apparent from the curve corresponding to the initial mass $M =$ 64 *amu*, clearly reveals itself for lower initial masses. From the

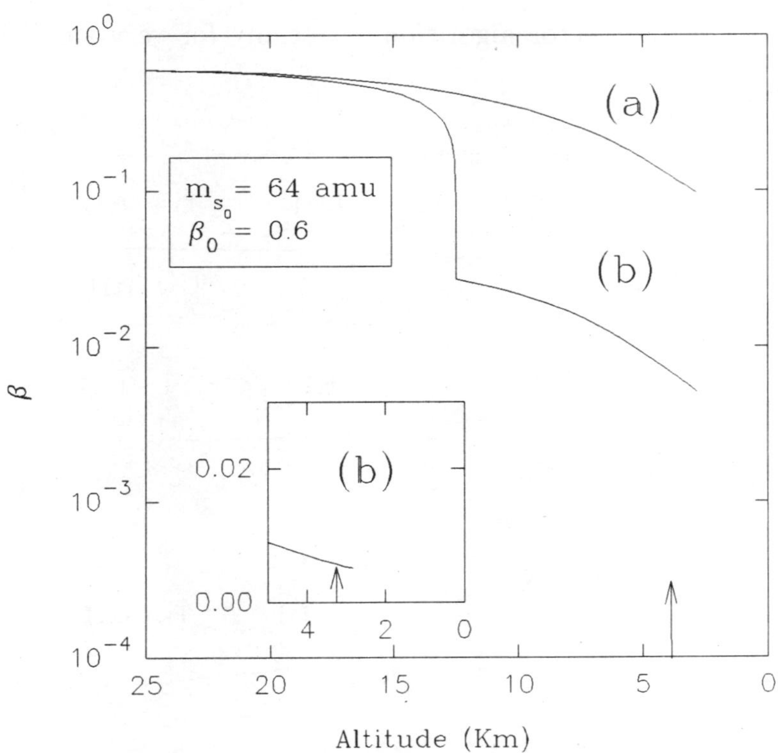

Figure 4: Variation of final β with altitude (a) for constant charge and without ionization loss and (b) including proton absorption as well as ionization loss

same figure, it can also be inferred that the value of β where the 'dip' occurs shifts to the left with increasing values of the initial mass.

Let us now consider a representative set of data with initial mass 64 *amu* and charge 2 for detailed discussion. The results for $\beta_0 = 0.6$ are shown in figures 4 and 5, where the variation of speed (β) and the energy of the strangelet with altitude are depicted. The sharp change seen at ~ 13 km corresponds to the onset of electron capture,

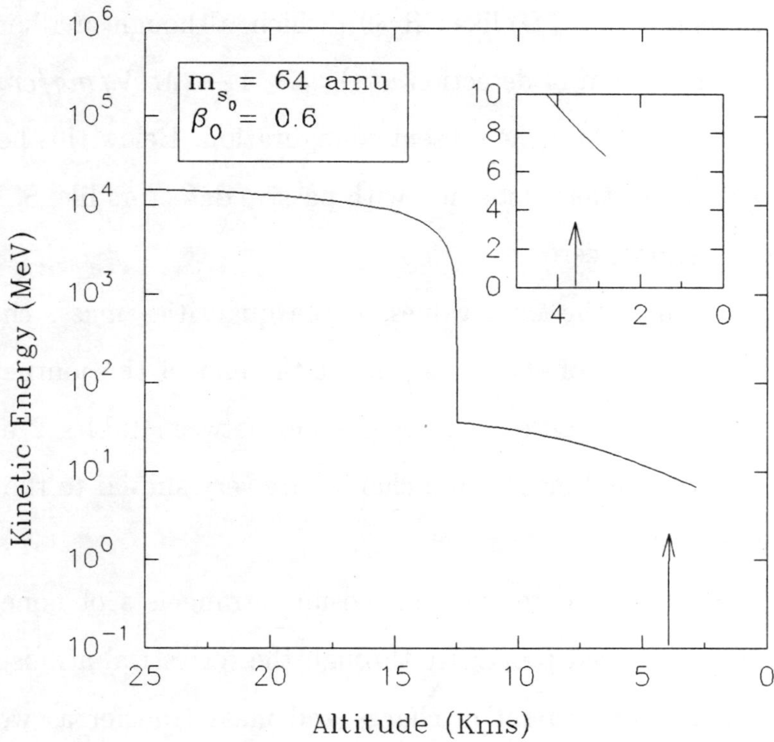

Figure 5: Variation of kinetic energy with altitude

which is handled phenomenologically through the effective Z_s. The insets of figures 4 and 5 show a zoomed-up view of the respective quantities near the endpoint of the journey. It is apparent from the figures that the ionization term reduces the overall energy and speed considerably from the nondissipative situation [36]. However, the zoomed-up insets in figs.4 and 5 show that the strangelets may have enough energy to be detectable at an altitude of 3.6 km from the sea level. For example, for the values of the initial quantities m_{s_o} and β_o shown here, the strangelet is left with a kinetic energy $\sim$8.5 MeV (corresponding to $\frac{dE}{dx} = 2.35\, MeV/mg/cm^2$ in a Solid State Nuclear

Track Detector (SSNTD) like CR-39), which, although small, is just above the threshold of detection $(\frac{dE}{dx})_{crit} \sim 1 - 2 MeV/mg/cm^2$ for $\beta < 10^{-2}$ in CR-39 for the present configuration. Below this height, the possibility of their detection with passive detectors like SSNTD reduces to almost zero.

Table 3 lists the final values of the quantities mass, charge, β, and the energy of the strangelet at the end of the journey for different initial velocities. A comparison between tables 2 and 3 shows that the final masses and charges are very similar to the ones found in cosmic ray events.

It thus stands to reason that cosmic strangelets of none-too-large size can indeed propagate through the terrestrial atmosphere and when proper account of charge and mass transfer as well as ionization loss is taken, they may indeed reach mountain altitudes. Therefore, a ground based large detector experiment set up at mountain altitudes and exposed for long times (two years or more) would have a good chance of detecting them. Such an effort is on in the eastern Himalayas and let me conclude by extending an invitation to the young students in the audience to consider joining this endeavour. Detection of strange quark matter in the cosmic rays would surely go a long way toward illuminating the nature of dark matter.

This work has been carried out over the past several years in collaboration with Jan-e Alam, Shibaji Banerjee, Pijushpani Bhattacharjee, Abhijit Bhattacharyya, Sanjay K. Ghosh, Amal Mazumdar, Bikash Sinha, Debapriyo Syam and Hiroshi Toki, whom it is my great pleasure to thank for their support and friendship.

Table 3: The final values, denoted with suffix l, are tabulated along with initial β (β_0)

β_0	m_{s_0}	m_l (amu)	q_l	$\beta_l \times (10^{-3})$	e_l (MeV)
	42	294.7	3	2.8	1.05
0.2	54	369.4	4	3.0	1.55
	60	415.8	4	3.0	1.80
	64	446.5	5	3.1	1.98
	42	246.4	6	4.9	2.84
0.4	54	359.5	8	4.7	3.73
	60	415.6	8	4.7	4.25
	64	452.0	9	4.6	4.63
	42	235.8	10	7.4	5.97
0.6	54	357.1	12	6.6	7.15
	60	416.0	13	6.4	7.87
	64	453.6	14	6.3	8.39
	42	236.4	12	8.6	8.16
0.7	54	359.1	14	7.6	9.59
	60	418.3	15	7.3	10.46
	64	456.3	16	7.2	11.11

References

[1] Wilkinson Microwave Anisotropy Probe data

[2] C. Alcock *et al, Nature (London)* **365**, 621 (1993)

[3] E. Aubourg *et al, Nature (London)* **365**, 623 (1993)

[4] B. Paczynski, *Astrophys. J.* **304**, 1 (1986)

[5] W. Sutherland , *Rev. Mod. Phys.* **71**, 421 (1999)

[6] E. Witten, *Phys. Rev.* **D30**, 272 (1984)

[7] J. Yang, M. S. Turner, G. Steigman, D. N. Schramm and K. A. Olive, *Astrophys. J.* **281**, 493 (1984)

[8] R. Schaeffer, P. Delbourgo-Salvador and J. Audouze, *Nature* **317**, 407 (1985)

[9] N. C. Rana, B. Datta, S. Raha and B. Sinha, *Phys. Lett.* **240B**, 175 (1990)

[10] P. Bhattacharjee, J. Alam, B. Sinha and S. Raha, *Phys. Rev.* **D48**, 4630 (1993)

[11] K. Sumiyoshi and T. Kajino, *Nucl. Phys.* **B24**, 80 (1991)

[12] J. Madsen and K. Riisager, *Phys. Lett.* **158B**, 208 (1985)

[13] J. Alam, S. Raha and B. Sinha, *Ap. J.* **513**, 572 (1999)

[14] K. Iso, H. Kodama and K. Sato, *Phys. Lett.* **B169**, 337 (1986)

[15] D. Stauffer, *Phys. Rep.* **54**, 1 (1979)

[16] A. Bhattacharyya, J. Alam, P. Roy, S. Sarkar, B. Sinha, S. Raha and P. Bhattacharjee, *Phys. Rev.* **D61**, 083509 (2000)

[17] H. Kodama, M. Sasaki, K. Sato, *Prog. Theo. Phys.* **68**, 1979 (1982)

[18] W. N. Cottingham, D. Kalafatis and R. Vinh Mau, *Phys. Rev. Lett.* **73**, 1328 (1994)

[19] K. Kajantie, *Phys. Lett.* **285B**, 331 (1992)

[20] P. J. E. Peebles, *The large scale structure of the universe*, Princeton Univ. Press, Princeton, NJ, 1980

[21] V. B. Bhatia, *A textbook on Astronomy and Astrophysics, with elements of Cosmology*, Narosa Publishing, New Delhi, 2001

[22] A. Chodos, R. L. Jaffe, K. Johnson, C. B. Thorn and V. F. Weisskopf,*Phys. Rev.* **D9**, 3471 (1974)

[23] S. Banerjee, S. K. Ghosh, and S. Raha, *J. Phys.* **G26**, L1 (2000)

[24] R. Narayan and M. Bartelmann, in: *Formation of structure in the universe* (A. Dekel and J. P. Ostriker, editors), Cambridge University Press, Cambridge, 1999

[25] J. Schaffner-Bielich, C. Greiner, A. Diener and H. Stöcker, *Phys. Rev.* **C55**, 3038 (1997); J. Schaffner-Bielich, *J. Phys. G : Nucl. Part. Phys.* **23**, 2107 (1997)

[26] Jes Madsen, *astro-ph/***9809032**; in : **Hadrons in Dense Matter and Hadrosynthesis**, *Lecture Notes in Physics*, Springer Verlag, Heidelberg, 1998

[27] M. Kasuya *et al.*, *Phys. Rev.* **D47**, 2153 (1993)

[28] M. Ichimura *et al.*, *Il Nuovo Cim.* **A106**, 843 (1993)

[29] P. B. Price *et al*, *Phys. Rev.* **D18**, 1382 (1978); T. Saito, *Proc. 24th ICRC, Rome* **1**, 898 (1995)

[30] O. Miyamura, *Proc. 24th ICRC Rome* **1**, 890 (1995)

[31] J. N. Capdeville, *Il Nuovo Cim.* **19C**, 623 (1996)

[32] G. Wilk and Z. Wlodarczyk, *J. Phys. G : Nucl. Part. Phys.* **22**, L105 (1996)

[33] G. Wilk and Z. Wlodarczyk, *Nucl. Phys. (Proc. Suppl.)* **52B**, 215 (1997)

[34] G. Wilk and Z. Wlodarczyk, *Heavy Ion Phys.* **4**, 395 (1996)

[35] E. Gadysz-Dziadus and Z. Wlodarczyk, *J. Phys. G : Nucl. Part. Phys.* **23**, 2057 (1996)

[36] S. Banerjee, S. K. Ghosh, S. Raha and D. Syam, *J. Phys. G : Nucl. Part. Phys.* **25**, L15 (1999)

[37] G. W. C. Kaye and T. H. Laby, *Tables of Physical and Chemical Constants and Some Mathematical Functions*, Longman, New York and London (1986)

[38] H. A. Bethe, *Ann. Phys.* **5**, 325 (1930); H. A. Bethe and J. Ashkin, in : *Experimental Nuclear Physics*, Vol. I (ed. E. Segrè), p. 166, John Wiley and Sons, New York and London, 1953

[39] Aa. Bohr, *Mat. Fys. Medd. Dan. Vid. Selsk.* **24**, No. 19 (1948)

[40] **Fundamental Formulas of Physics, Vol. 2**, *Donald H. Menzel (Ed)*, p. 560, Dover Publications, New York (1960)

Physics of Solids, Nuclei and Particles
Editor: R. Sahu

Structure of Hadrons

S.C. Phatak

Institute of Physics, Bhubaneswar 751 005, India

Abstract

In this talk we describe the present understanding of the structure of hadrons in terms of the quantum chromodynamics. In particular, we discuss the implication of the hadron structure on the understanding of nuclear physics. We conclude that, broadly speaking, the present understanding of nuclear physics in terms of point-like nucleons interacting via meson exchange potentials holds. However, it is still not clear how the meson exchange potentials arise from quantum chromodynamics.

1 Introduction

Classically, in nuclear physics on assumes that the nuclei consist of non-relativistic point-like nucleons interacting via two-nucleon potentials. It is understood that the two-nucleon exchange potentials are produced by exchanges of various mesons. Using this picture one is able to explain a vast amount of data. For example, one can understand the properties of two nucleon bound state (the deureton) as well as the nucleon-nucleon scattering data from meson exchange potentials. Similarly, using many body theories, such as Hartree-Fock and its variants, one can understand the properties of nuclei. A detailed exposition of this is found in nuclear physics texts.

On the other hand, we now know that the basic theory of strong interaction, which is supposed to provide the structure for the nuclear physics, is quantum chromodynamics (QCD). In QCD, the nucleons are not elementary objects and they consist of elmentary objects called quarks. There is also a sufficient evidence to show that the nucleons are extended particles. One may, therefore, think whether the the classical description of nuclear physics in terms of point nucleons interacting through exchanges of mesons still remains valid or whether one should have an entirely different description of nuclear physics in terms of quarks and QCD. We shall be addressing this problem in this talk.

To begin with we shall see how one learnt about the structure of hadrons (nucleons in particular) and how the theory of QCD was developed. We shall then describe the saliant features of QCD and finally, we shall discuss how these features give rise to the classical picture of nuclear physics.

2 Structure of Nucleon : Small Momentum Transfers

As discussed in the introduction, one considers nucleons to be point particles interacting through two-body interaction. Although this model works very well, it is known for a long time that the nucleons are not really point particles. The first evidence for the finite extent of the nucleons comes from the magnetic moment of the nucleon. We know that the nucleons are spin-1/2 particles. Also, there is a very successful theory of spin-1/2 particles given by Dirac and in this theory, the magnetic moment of spin-1/2 particles arises naturally. Classically, for charged but point particles, we do not expect magnetic moment because, essentially the magnetic moment arises from the circular motion of charge (or circular currents). But Dirac's theory showed that the charged spin-1/2 particle has magnetic moment $\mu = \frac{e\hbar}{2mc}$ where e is the charge of the particle, m is its mass, $\hbar$ is the Plank's constant and c is the velocity of light.

The magnetic moment of the nucleon is given by this expression[2]. Clearly, the Dirac's theory implies that a neutral spin-1/2 particle shuld not have any magnetic moment.

The measured magnetic moments of proton and neutron differ from the value given above. The measuerd values of proton and neutron magnetic moments are $2.79\cdots$ and $-1.91\cdots$ nuclear magnetons respectively. Here the nuclear magneton is $\frac{e\hbar}{2Mc}$ where M is the nucleon mass. According to the Dirac theory, these values should be 1 and 0 nuclear magnetons respectively. So, this means that either the Dirac theory is not applicable to nucleons or the nucleons are not point particles. One prefers the latter explanation because Dirac theory has been quite successful otherwise. Thus, the nuclear magnetic moments give an indication that the nucleons are extended objects. For example, one can understand the negative value of neutron magnetic moment if one thinks that the neutron consists of a positively charged core with a negatively charged cloud circling around it. However, the magnetic moment is a static property and therefore it cannot tell much about the details of the nucleon structure. All that it says is that the nucleons are not point particles.

A more direct evidence for the structure of nucleons comes from electron scattering experiments. Let us consider the elestic scattering of electrons from a point particle. This is the classical Rutherford scattering. The scattering cross section for this process is given by

$$\frac{d\sigma}{d\Omega} \propto \left[\frac{qe}{q^2}\right]^2 = \left[\frac{qe}{4k^2\sin^2(\theta/2)}\right]^2 \tag{1}$$

where q and e are the charges of the point particle and the electron respectively, k is the incident relative momentum of electron-particle system, q is the momentum transfer and θ is the scattering angle. Here we have ignored the contribution due to magnetic moments. Note that the potential between the electron and a point charge is $V(r) = \frac{qe}{r}$ where r is the distance between the electron and the

---

[2]Actually, the electronic magnetic moment differs slightly from the value given above and this difference is well explained by the theory of quantum electrodynamics. Here we are not considering this difference

point particle. So the Fourier transform of this potential is

$$V(q) \;=\; \int d^3 r \, e^{i\vec{q}\cdot\vec{r}} V(r)$$

$$\propto \; \frac{qe}{q^2} \tag{2}$$

Thus the quantity in the square brackets above is the Fourier transform of the Coulomb potential between electron and the point charge.

In fact, in Born approximation, the scattering cross section is proportional to the square of the Fourier transform of the interaction potential. And, because the Coulomb interaction is weak and can be treated perturbatively, the Born approximation works very well for electron scattering [3]. When on replaces the point charge by a charge distribution $\rho(r)$, the potential between the electron and the charge distribution is given by

$$V(r) = e \int d^3 r' \frac{\rho(r')}{|\vec{r} - \vec{r}'|} \tag{3}$$

and the Fourier transform of this potential is

$$V(q) \propto \frac{qe}{q^2} F(q) \tag{4}$$

where $F(q)$ is the Fourier transform of the charge distribution $\rho(r)$. Thus, if the Born approximation is valid (which is the case for Coulomb scattering since the Coulomb potential is weak) the ratio of the scattering cross sections for charge distribution and point charge gives the square of the Fourier transform of the charge distribution. One can therefore compute the charge distribution by taking the Fourier transform of the square root of this ratio. In other words, electron scattering allows us to determine the charge distribution of an extended object.

Unfortunately, things are not as simple as that. Major problem is, to determine the density we have to know the ratio of cross sections for all values of momentum transfer ($q = \sqrt{(\vec{k} - \vec{k}')^2} =$

---

[3]Of course in actual calculation one solves the scattering problem exactly and includes the magnetic moment contributions. But the arguments given here are valid even when these effects are incorporated.

$2k\sin(\theta/2)$. So, for given value of electron energy, the ratio is known for a restricted values of q. This means that the charge distribution can only be determined approximately. At small electron energies, one can measure the form factor for a very restricted values of q. Thus, one can determine the value of the form factor at zero momentum transfer and its derivative. One can show that this gives the total charge and root mean square of the charge distribution.

Electron scattering experiments have been done and the mean square charge distributions for proton and neutron have been measured to be $\sim 0.67 fm^2$ and $\sim -0.12 fm^2$ respectively. So, again one finds that the nucleons are not point particles. If you assume that the proton has a uniform spherical charge distribution, the above value implies that the charge radius of proton is about 1 fm. Note that the average distance between nucleons in a nucleus is about 2 fm. So, the nucleons in a nucleus are very close to touching each other. Also note that the neutron square charge radius is negative. This implies that the neutron charge distribution can be thught to be a positively charged core surrounded by a negatively charged cloud. This ties up with the magnetic moment value of neutron.

Note that the fact that nucleons have anomolous magnetic moments and nonzero charge radius means that the nucleons are extended objects. It does not imply that they are composite objects made out of some more fundamental objects. At these energies or momentum transfers, electron scattering cannot resolve the details of the nucleon structure. Later we shall consider electron scattering at higher energies and that will tell us about the constituents of the hadrons. But before that we shall discuss another development which lead to the postulate of quark model.

3   The Quark Model

By early sixties, a large number of particles and resonances were discovered in pion-nucleon scattering experiments. So, calling all of them as elementary particles did not make sense. One doesn't wan too many elementary particles. The situation was similar to the case of atoms before the discovery of proton, electron and neutron. Atoms were thought to be the building blocks of matter but having close to hundred species of atoms, which were elementary did not seem good. But, in case of atomic physics, the electron, proton and neutron were discovered first and then models or theories of construction of atoms were developed. In case of hadrons, the story developed the other way. It was found that the particles and resonances discovered in pion scattering experiments can be classified into representations of SU(3) group.

Let me illustrate this by consdering an example of isospin. We know that neutron and proton have almost same mass. Their mas differs by couple of MeV and this difference is about 0.2 % of their mass. There are three π mesons having masses very close to each other. So, one can think that neutron-proton pair and three pions are multiplets of same entity. Just like a particle having spin projection $+1/2$ and $-1/2$ are really same object but have it's spin projection is different. This can be extended to nuclei as well. For example, ^3H and ^3He or ^{13}C and ^{13}N form pairs having similar binding energies, similar excitation spectrum and so on. This can be explained by assuming that these objects constitute different representations of SU(2) group. This group is called isospin group. The fundamental representation of the SU(2) group is a doublet (very much like the two projections of spin-1/2 particle). Of course the SU(2) of isospin is an approximate symmetry of nature. The electromagnetic interaction does not obey isospin since proton has one unit of charge and nuetron is neutral. But the strong interactions seem to obey it to a great extent. So, SU(2) of isospin helps in the classification of particles (elementary particles as well as nuclei). Further, the energy (or mass) splitting within isospin multiplet can be understood.

The fact that the particles can be classified according to SU(3) group implies that the SU(3) symmetry is at least an approximate symmetry of strong interactions. Actually, this SU(3) has a SU(2) subgroup which isidentified with the isospin. This subgroup seems to be a good symmetry of strong interaction but the extension (which is required to include strangeness) seems to be badly broken. Thus, mas difference between two nucleons, three Σ's etc is very small (few MeV's) but the mass difference between strangeness zero, -1 and -2 baryons is hundreds of MeV. But still, the classification under SU(3) works very well.

Just like the fundamental representation of SU(2) is a doublet, the funamental representation of SU(3) is a triplet. Since any other representations of SU(3) can be built from the fundamental representation, one can postulate three particles which constitute the fundamental representation and build strongly interacting particles (baryons and mesons) out of these. This is exactly what Gell-Mann and Zweig did in 1964. These postulated particles were called quarks and the analysis implied that the baryons are constructed from three quarks and the mesons are constructed from a quark-antiquark pair.

It is interesting to note that GellMann thought that the quarks are ficticious particles and don't have any physical meaning. One reason why he may have thought so was that no such particles were observed experimentally. Further, these particles rquired fractional charges (2/3 for u quark and -1/3 for d and s quarks) and there was absolutely no evidence for fractional charges. It was only after the deep inelastic scattering experiments it was realised that the nucleons (or hadrons generally) have constituents within them. We shall consider these aspects now.

4   Deep Inelastic Electron Scattering

Let us come back to the electron scattering again. We saw that, qualitatively the electron scattering cross sections can be expressed as a product of scattering from a point particle times the square of the Fourier transform of the charge distribution. The latter is also called as form factor. Generally, for any object having finite extent, it's form factor decreases rapidly as the momentum transfer is increased. Generally in a scattering experiment, the target (the nucleon in this case) may be left in the ground state (elastic scattering) or may be left in an excited state (inelastic scattering). So, when one looks at the excitation spectrum (that is the scattering cross section as a function of outgoing projectile energy), one would observe peaks corrsponding to elastic scattering, excitation to various excited states or resonances and so on. The finite size of hadrons means that the strength of these excitations (the height of the peaks) should decrease when the momentum transfer is increased. This is because of the decrease of the form factors with the increase in the momentum transfer.

Now, if the target has elementary constituents, the electron is actually scattering from these constituents. When one is considering elastic scattering or stattering to resonances, the scattered constituent interacts with the rest of the constituents of the nucleon to form the final state nucleon or the resonance. At sufficiently large momentum transfers, this is difficult because the scatterd constituent carries large momentum transfer and it is difficult to share such large momentum transfer between other constituents efficiently. This is because of the size of the nucleon or resonances. Thus the cross sections for such processes drops at large momentum transfers. But at such large momentum transfers, the reaction is so fast that the effect of other constituents on the scattered constituent can be ignored. This is called quasi-free scattering or impulse approximation. Thus, the scattering is equivalent to the scattering from the constituent (as if the other constituents didn't exist). A consequence of this is that the strength of the excitation function for such a process does not depend on the momentum transfered to the scattered constituent (provided the constituent is point-like. Such a quasi-free scattering gives rise to a broad bump in the cross section. And the structure of the bump does not depend on the incident energy of the electron. What this means is that by looking at the excitation function (or the electron scattering cross section at a fixed angle and as a funcion of outgoing electron energy) one will be able to decide whether the scattering is taking place from the nucleon as a whole or from the constituents of the nucleon. As mentioned above, this argument is valid if the constituents don't have any structure and are point-like.

It turns out that when one includes relativistic kinematics, the excitation function depends on the quantity $x = Q^2/2M\nu$ where Q is the four momentum transfer, M is the nucleon mass and ν is the energy loss of the enectron in lab frame. It does not depend on the four momentum transfer and the energy loss independently. This property is called Bjoken scaling and the quantity x can be identified with the momentum fraction of the nucleon carried by the constituent. The verification of Bjorken scaling established that the nucleon (and other hadrons) have point-like constituents inside.

So, the deep inelastic scattering experiments establish that the hadrons contain elementary point-like objects withing them. Although we shall not discuss it here, it has also been established that these partons have 1/2 units of angular momentum. Thus, these can be identified with the quarks of Gell-Mann and Zweig. These experiments have also established that there are other constituents of hadrons which are charge neutral.

5   The Quantum Chromodynamics

From preceeding discussion we have established that nucleons (and hadrons in general) are composite and their constitients are the quarks of GellMann and Zweig. Although these are not observed in isolation, these are seen indirectly in deep inelastic scattering experiments. The question then is, why are these not liberated from hadrons (as one can liberate electrons from atoms) and observe them in isolation. Further, there are some problems with the statistics. In nutshell, quarks are Fermions and one needs three of them to form baryons. But in some baryonic states, like the Δ resonance, the quark wave function has to be symmetric. Also, the spin-flavour wave function of baryon octet (the nucleon, Σ etc), according to SU(3) model of GellMann, is symmetric and one also expects the space wave function to be symmetric. This is not consistant with the spin-statistics theorem.

To overcome the problem mentioned above and many other problems, quarks were attributed with an internal degree of freedom called colour. This is much like the isospin degree of fredom of nuclear physics. However, the colour symmetry needs to be an exact symmetry, unlike the isospin which we know is approximate. Another difference between the isospin and colour symmetry is that the fundamental representation of the isospin is a doublet where as that of colour is a triplet. Thus, mathematically, the isospin symmetry is described by the SU(2) group where as the colour symmetry is described by the SU(3) group. In that sense, the colour symmetry is similar to the SU(3) symmetry introduced by GellMann and Neeman. The posulate of colour degree of freedom is further verified by a number of experimental results, such as π^0 decay, ratio of e^+e^- annihilation cross section to hadrons and $\mu^+\mu^-$ and so on. We cannot go into the details of these experiments for lack of time. An interesting development, taking advantage of the exact colour symmetry, one can introduce a theory for strong interaction. This has been done in analogy with the derivation of electromagnetic theory by using the gauge principle. Again we will not go into the details of how the theory of strong interaction (now called as quantum chromodynamics (QCD)) was developed but state some important features of the theory.

1. Although the QCD is developed in analogy with the QED, there are several fundamental difference between them. These essentially arise from the fact that the underlying group for QED is U(1) where as for QCD it is SU(3). This leads to following differences:

   - There is one gauge boson (the photon) in QED where as there are eight guage bosons (the gluons) in QCD.

   - The photons do not carry electric charge where as the gluons carry colour charge. Thus, the photons cannot interact with each other by exchanging aphoton but gluons do interact by enchanging a gluon (or two gluons).

   - Renormalization group calculations show that the electromagnetic interaction becomes stronger as the momentum transfer is increased. In case of the strong interaction, the situation is opposite. Conversely, the strong interaction becomes stronger at smaller momentum transfers.

2. The consequence of the renormalization group result is that one can perform large momentum strong interaction calculations perturbatively and, indeed, one obtains a good agreement with the data. This property of QCD is called the asymptotic freedom.

3. Conversely, the QCD interaction becomes strong at small momntum transfers or at large distance scales. This property is called confinement. What it means is that the only strongly interacting objects which can be directly seen are those which are colour neutral or which are singlets of colour SU(3) group. Thus, quarks and gluons, which carry colour charge, cannot be directly seen in experiments. But a colour singlet combination can be. In other words, all the observed hadrons, baryons and mesons, are singlets of colour SU(3) group. Indeed, this is observed.

4. The masses of u and d quarks are observed to be small, in comparison with the typical masses of hadrons. The u and d quark masses are about 10 MeV where as the lightest hadron is π meson which is 140 MeV and other baryons masses are larger that 500 MeV. This small quark mass leads to an approximate symmetry called chiral symmetry. One can assume that the QCD has this symmety. Mathematical details show that this symmetry can, in principle, manifest in two different ways. One is called the Wigner realization in which the states particles) are parity doublets. The other is called Goldstone realization, in which the symmetry is spontaneously broken and this breaking gives rise to a massless Goldstone boson.

5. It turns out that one does not have the Wigner realization in the real world since we do not find pairs of opposite parity particles. But it is realized in the Goldstone mode and one can argue that the pion is the goldstone boson produced due to the spontaneous symmetry breaking.

The preceeding discussion shows that the theory of strong interaction, the QCD, has certain saliant features, like asymptotic freedom, confinement and chiral symmetry which is realized in Goldstone mode. The fundamental question now arises is, do we need to change the basis of nuclear physics. That is, since the interaction of nuclear physics is the strong interaction. Now that the strong interaction is considered to be arising from the QCD, is it now necessary to start from the QCD to describe the nuclear physics. That is, is it now necessary to change the basic picture of nuclear physics and describe the nuclei, their properties, their interactions in terms of the theory of QCD. Or, is it possible to adopt a two-step approach in which one deduces the nucleon-nucleon interaction from QCD and uses this interaction to describe the nuclei and their interactions. This is a very important question because if the latter explanation is not valid, we will have to re-learn the nuclear physics in terms of quarks, gluons and QCD. In the following section, we shall argue that the former approach is, still by and large valid and we don't need to re-write the nuclear physics text books.

6 Nuclear Physics - Post-QCD Understanding

AS discussed in the previous section, we shall now consider the arguments for and against attempting to describe the nuclear physics in terms of quarks and QCD.

1. We have seen that the QCD coupling constant becomes large when the momentum transfer is small. Or equivalently, at large separations the QCD coupling constant is large. This happens for distance scales of the order of fermi or less. Thus, when one is considering nuclear physics, the corresponding QCD coupling constant is large. This means standard methods like perturbation theory are not applicable and therefore it is practically impossible to describe the physics at distance scales of the order of fermi or more in terms of the QCD. In other words, one cannot describe nuclei and their interaction in terms of quarks and gluons as one describes atoms in terms of nuclei and electrons and Coulomb interaction. One needs to devise non-perturbative techniques which can be applied in such a situation. Unfortunately, such techniques don't exist at the moment.

2. From nuclear physics point of view, meson exchange picture of strong interactions is capable of explaining a vast amount of data on nucleon-nucleon scattering, deuteron bound state,

meson-nucleon scattering as well as producion of baryon and meson resonances. The same picture is able to explain properties of nuclei using many body theories such as Hartree-Fock. Obviously, any theory developed from the QCD should be able to explain all this wealth of data. It is unlikely that an entirely new theory, which will replace the current understanding of nuclear physics will emerge from the QCD.

3. It is therefore likely that the understanding of nuclear physics from QCD will be in two steps. The first step would be to understand the meson exchange picture from QCD and the second step, about which we know a lot already, is to explain the nuclear physics in terms of the meson exchange picture.

4. Now I shall argue that a meson exchange picture does emerge from the QCD if we take the essential features of the QCD into account. We have seen that the three important properties of QCD are asymptotic freedom, confinement and chiral symmetry. The confinement implies that only colour neutral objects or objects which are singlets of colour SU(3) group are observable. Indeed, all the observed baryons and mesons are singlets of colour. The size of hadrons is about a fermi, so that is consistant with confinement. In fact, there are a number of QCD-based models (like MIT bag model) which incorporate the idea of confinement nicely. And these are able to explain the spectroscopy of hadrons reasonably well. Where these models fail is in describing the interactions between hadrons (and therefore, in describing nuclear physics). This is because these models do not have the long range interaction between hadrons.

5. We have also argued that the QCD respects chiral invariance approximately. This is because the masses of u and d quarks are small. However, the chiral symmetry does not manifest itself in Wigner mode because hadronic states are not parity doublets. For example, nucleons have positive parity and we don't have a negative parity baryons having masses close to nucleon mass. This means the symmetry must manifest itself

in Goldstone mode. This means that the symmetry is broken spontaneously and this gives rise to a Goldstone boson having zero mass. Now the chiral symmetry is not an exact symmetry and therefore the Goldstone boson is not massless but has a small mass. This Goldstone boson is the pion. It is interesting to find that the spontaneous breaking of chiral symmetry not only leads to the identification of pion as the Goldstone boson but it also gives us the interaction of pion with other hadrons. One can extend the QCD-based models (such as the bag model) to incorporate this aspect of QCD. The nice thing is that such models automatically give the coupling of pion to the hadrons and the end result is we have a meson theory of interaction between hadrons.

The statements given above can be worked out in details and one can build models of hadrons which incorporate the two most important aspects of QCD; confinement and chiral invariance. These models, although not derived from QCD, one can argue that QCD is expected to provide a foundation for these models. One of the consequences of these models is that the interaction between hadrons, and between nucleons in particular, is already present. This interaction arises because of incorporation of the chiral invariance. Basically, the chiral invariance gives rise to coupling of pion mesons with the hadrons and because of this coupling it is possible to exchange pions between hadrons to generate the interaction between hadrons. It must be emphasized that the long range part of nuclear interaction is generated by the exchange of pion and this is automatically generated in these chiral models.

Thus it appears that meson exchange interactions arise naturally in QCD because of the approximate chiral invariance. Taking advantage of the chiral invariance it is possible to devise models which generate meson coupling to hadrons. Using these models one essentially recovers the classical picture of nuclear physics, that is the nuclear physics can be described in terms of point-like nucleons interacting by exchanges of mesons. What remains to be done is to 'derive' the meson hadron coupling starting from QCD.

7   References

We have not given any references in the text. The material presented in this article is collected from various sources. Some of the references for further reading are listed below.

1. M. A. Preston and R. Bhaduri, "Structure of The Nucleus" (Addison-Wesley, Reading, MA, 75)

2. F. E. Close, "Introduction to Quarks and Partons" (Academic, London, 79)

3. A. W. Thomas, Adv. Nucl. Phys. 13, 1(84)

4. F. Halzen and A. D. Martin, "Quarks and Leptons" (Wiley, N.Y., 84)

5. T. D. Lee, "Particle Physics and Introduction to Field Theory", (Harwood Academic, N.Y., 81)